An R Companion to
Linear Statistical Models

An R Companion to Linear Statistical Models

Christopher Hay-Jahans

CRC Press
Taylor & Francis Group
Boca Raton London New York

CRC Press is an imprint of the
Taylor & Francis Group, an **informa** business

A CHAPMAN & HALL BOOK

CRC Press
Taylor & Francis Group
6000 Broken Sound Parkway NW, Suite 300
Boca Raton, FL 33487-2742

First issued in paperback 2017

ISBN-13: 978-1-4398-7365-6 (hbk)
ISBN-13: 978-1-138-11603-0 (pbk)

Library of Congress Cataloging-in-Publication Data

Hay-Jahans, Christopher.
    An R companion to linear statistical models / Christopher Hay-Jahans.
      p. cm.
    Summary: "Focusing on user-developed programming, An R Companion to Linear Statistical Models serves two audiences: Those who are familiar with the theory and applications of linear statistical models and wish to learn or enhance their skills in R; and those who are enrolled in an R-based course on regression and analysis of variance. For those who have never used R, the book begins with a self-contained introduction to R that lays the foundation for later chapters.This book includes extensive and carefully explained examples of how to write programs using the R programming language. These examples cover methods used for linear regression and designed experiments with up to two fixed-effects factors, including blocking variables and covariates. It also demonstrates applications of several pre-packaged functions for complex computational procedures. "-- Provided by publisher.
    Includes bibliographical references and index.
    ISBN 978-1-4398-7365-6 (hardback)
    1. Linear models (Statistics) 2. R (Computer program language) I. Title.

QA279.H39 2011
519.50285'5133--dc23
                                        2011039833

Visit the Taylor & Francis Web site at
http://www.taylorandfrancis.com

and the CRC Press Web site at
http://www.crcpress.com

To Joan, Phillip, Sandra, and Magil

Contents

Preface   xv

I Background   1

1 Getting Started   3
  1.1   Introduction   3
  1.2   Starting up R   4
  1.3   Searching for Help   7
  1.4   Managing Objects in the Workspace 10
  1.5   Installing and Loading Packages from CRAN 12
  1.6   Attaching R Objects 16
  1.7   Saving Graphics Images from R 17
  1.8   Viewing and Saving Session History 18
  1.9   Citing R and Packages from CRAN 19
  1.10 The R Script Editor 20

2 Working with Numbers   23
  2.1   Introduction 23
  2.2   Elementary Operators and Functions 23
       2.2.1   Elementary arithmetic operations 24
       2.2.2   Common mathematical functions 24
       2.2.3   Other miscellaneous operators and functions 25
       2.2.4   Some cautions on computations 26
  2.3   Sequences of Numbers 27
       2.3.1   Arithmetic sequences 27
       2.3.2   Customized sequences 28
       2.3.3   Summations of sequences 28
  2.4   Common Probability Distributions 29
       2.4.1   Normal distributions 29
       2.4.2   t-distributions 30
       2.4.3   F-distributions 31
  2.5   User Defined Functions 32

3 Working with Data Structures   **35**

3.1   Introduction   35

3.2   Naming and Initializing Data Structures   36

3.3   Classifications of Data within Data Structures   38

3.4   Basics with Univariate Data   40

    3.4.1   Accessing the contents of univariate datasets   40

    3.4.2   Sorting univariate datasets   41

    3.4.3   Computations with univariate datasets   42

    3.4.4   Special vector operations   44

    3.4.5   Importing univariate datasets from external files . . .   47

3.5   Basics with Multivariate Data   49

    3.5.1   Accessing the contents of multivariate datasets   49

    3.5.2   Computations with multivariate datasets   51

3.6   Descriptive Statistics   52

    3.6.1   Univariate datasets   53

    3.6.2   Multivariate datasets   54

3.7   For the Curious   55

    3.7.1   Constructing a grouped frequency distribution   56

    3.7.2   Using a sink file to save output   57

4 Basic Plotting Functions   **59**

4.1   Introduction   59

4.2   The Graphics Window   61

4.3   Boxplots   61

4.4   Histograms   63

4.5   Density Histograms and Normal Curves   67

4.6   Stripcharts   69

4.7   QQ Normal Probability Plots   70

4.8   Half-Normal Plots   72

4.9   Time-Series Plots   73

4.10   Scatterplots   75

4.11   Matrix Scatterplots   76

4.12   Bells and Whistles   76

4.13   For the Curious   78

    4.13.1   QQ normal probability plots from scratch   78

    4.13.2   Half-normal plots from scratch   79

    4.13.3   Interpreting QQ normal probability plots   80

5 Automating Flow in Programs   **83**

5.1   Introduction   83

5.2   Logical Variables, Operators, and Statements   83

    5.2.1   Initializing logical variables   84

    5.2.2   Logical operators   84

    5.2.3   Relational operators   85

5.3   Conditional Statements   87

	5.3.1	If-stop	87
	5.3.2	If-then	88
	5.3.3	If-then-else	88
	5.3.4	The ifelse function	89
5.4	Loops	. . . .	90
	5.4.1	For-loops	90
	5.4.2	While-loops	91
5.5	Programming Examples		91
	5.5.1	Numerical solution of an equation	92
	5.5.2	A snazzy histogram	93
5.6	Some Programming Tips		96

II Linear Regression Models

99

6 Simple Linear Regression

101

6.1	Introduction	101
6.2	Exploratory Data Analysis	102
6.3	Model Construction and Fit	104
6.4	Diagnostics	106

	6.4.1	The constant variance assumption	107
		6.4.1.1 The F-test for two population variances . . .	108
		6.4.1.2 The Brown–Forsyth test	109
	6.4.2	The normality assumption	110
		6.4.2.1 QQ normal probability correlation coefficient test	111
		6.4.2.2 Shapiro–Wilk test	112
	6.4.3	The independence assumption	112
	6.4.4	The presence and influence of outliers	113
6.5	Estimating Regression Parameters		116
	6.5.1	One-at-a-time intervals	116
	6.5.2	Simultaneous intervals	118
6.6	Confidence Intervals for the Mean Response		119
	6.6.1	One-at-a-time t-intervals	119
	6.6.2	Simultaneous t-intervals	121
	6.6.3	Simultaneous F-intervals	121
	6.6.4	Confidence bands	123
6.7	Prediction Intervals for New Observations		125
	6.7.1	t-interval for a single new response	125
	6.7.2	Simultaneous t-intervals	126
	6.7.3	Simultaneous F-intervals	127
	6.7.4	Prediction bands	128
6.8	For the Curious		129
	6.8.1	Producing simulated data	129
	6.8.2	The function ci.plot	130
	6.8.3	Brown–Forsyth test revisited	130

7 Simple Remedies for Simple Regression   **133**

7.1   Introduction 133

7.2   Improving Fit 134

     7.2.1   Transforming explanatory variables 134

     7.2.2   Transforming explanatory and response variables . . . 137

7.3   Normalizing Transformations 138

     7.3.1   Transformations suggested by the theory 138

     7.3.2   Power (or log) transformations by observation 139

     7.3.3   The Box–Cox procedure 141

7.4   Variance Stabilizing Transformations 144

     7.4.1   Power transformations 144

     7.4.2   The arcSine transformation 146

7.5   Polynomial Regression 147

7.6   Piecewise Defined Models 149

     7.6.1   Subset regression 150

     7.6.2   Continuous piecewise regression 151

     7.6.3   A three-piece example 152

7.7   Introducing Categorical Variables 155

     7.7.1   Parallel straight-line models 156

     7.7.2   Non-parallel straight-line models 159

7.8   For the Curious 161

     7.8.1   The Box–Cox procedure revisited 162

     7.8.2   Inserting mathematical annotation in figures 164

     7.8.3   The split function 165

8 Multiple Linear Regression   **167**

8.1   Introduction 167

8.2   Exploratory Data Analysis 169

8.3   Model Construction and Fit 171

8.4   Diagnostics 173

     8.4.1   The constant variance assumption 174

         8.4.1.1   F-test for two population variances 176

         8.4.1.2   The Brown–Forsyth test 177

     8.4.2   The normality assumption 177

         8.4.2.1   QQ normal probability correlation coefficient test 178

         8.4.2.2   Shapiro–Wilk test 178

     8.4.3   The independence assumption 178

     8.4.4   The presence and influence of outliers 179

         8.4.4.1   Outlying observed responses 179

         8.4.4.2   Outliers in the domain space 181

         8.4.4.3   Influence of outliers on corresponding fitted values 181

         8.4.4.4   Influence of outliers on all fitted values . . . 182

         8.4.4.5   Influence of outliers on parameter estimates . 182

	8.4.4.6	Quick influence analysis	184
8.5	Estimating Regression Parameters	186
	8.5.1	One-at-a-time t-intervals	186
	8.5.2	Simultaneous t-intervals	186
	8.5.3	Simultaneous F-intervals	187
8.6	Confidence Intervals for the Mean Response	188
	8.6.1	One-at-a-time t-intervals	188
	8.6.2	Simultaneous t-intervals	189
	8.6.3	Simultaneous F-intervals	190
8.7	Prediction Intervals for New Observations	192
	8.7.1	t-interval for a single new response	192
	8.7.2	Simultaneous t-intervals	192
	8.7.3	Simultaneous F-intervals	194
8.8	For the Curious	195
	8.8.1	Fitting and testing a model from scratch	195
	8.8.2	Joint confidence regions for regression parameters . . .	197
	8.8.3	Confidence regions for the mean response	199
	8.8.4	Prediction regions	202

9 Additional Diagnostics for Multiple Regression — **203**
9.1	Introduction	203
9.2	Detection of Structural Violations	204
	9.2.1	Matrix scatterplots	204
	9.2.2	Partial residual plots	204
	9.2.3	Testing for lack of fit	205
9.3	Diagnosing Multicollinearity	208
	9.3.1	Correlation coefficients	209
	9.3.2	Variance inflation factors	210
9.4	Variable Selection	211
	9.4.1	Contribution significance of individual variables	211
	9.4.2	Contribution significance of groups of variables	211
	9.4.3	The drop1 function	213
	9.4.4	The add1 function	213
	9.4.5	Stepwise selection algorithms	214
9.5	Model Selection Criteria	215
	9.5.1	Mean square error and residual sum of squares	218
	9.5.2	Coefficient of determination	218
	9.5.3	Adjusted coefficient of determination	219
	9.5.4	Mallow's statistic	219
	9.5.5	Akaike and Bayesian information criteria	220
	9.5.6	PRESS statistic	221
9.6	For the Curious	222
	9.6.1	More on multicollinearity	222
		9.6.1.1	Background discussion	222
		9.6.1.2	The condition number	224

    9.6.1.3 Condition indices 225
   9.6.2 Variance proportions 226
   9.6.3 Model selection criteria revisited 228

10 Simple Remedies for Multiple Regression   **231**
 10.1 Introduction 231
 10.2 Improving Fit 232
   10.2.1 Power transformations of explanatory variables 232
   10.2.2 Transforming response and explanatory variables . . . 235
 10.3 Normalizing Transformations 238
 10.4 Variance Stabilizing Transformations 239
 10.5 Polynomial Regression 239
 10.6 Adding New Explanatory Variables 241
 10.7 What if None of the Simple Remedies Help? 242
   10.7.1 Variance issues 242
   10.7.2 Outlier issues 242
   10.7.3 Multicollinearity issues 242
   10.7.4 Autocorrelation issues 243
   10.7.5 Discrete response matters 243
   10.7.6 Distributional issues in general 243
 10.8 For the Curious: Box–Tidwell Revisited 243

III Linear Models with Fixed-Effects Factors **249**

11 One-Factor Models   **251**
 11.1 Introduction 251
 11.2 Exploratory Data Analysis 253
 11.3 Model Construction and Fit 254
 11.4 Diagnostics 257
   11.4.1 The constant variance assumption 257
   11.4.2 The normality assumption 258
   11.4.3 The independence assumption 259
   11.4.4 The presence and influence of outliers 260
 11.5 Pairwise Comparisons of Treatment Effects 261
   11.5.1 With a control treament 261
   11.5.2 Without a control treament 262
     11.5.2.1 Single pairwise t-test 262
     11.5.2.2 Simultaneous pairwise t-tests 263
     11.5.2.3 Tukey's HSD procedure 263
 11.6 Testing General Contrasts 265
 11.7 Alternative Variable Coding Schemes 267
   11.7.1 Treatment means model 267
   11.7.2 Treatment effects model 268
     11.7.2.1 Weighted mean condition 268
     11.7.2.2 Unweighted mean condition 270

    11.7.3 Diagnostics and pairwise comparisons 271
 11.8 For the Curious 271
    11.8.1 Interval Estimates of Treatment Means 271
       11.8.1.1 t-intervals 272
       11.8.1.2 Scheffe F-intervals 272
    11.8.2 Tukey–Kramer pairwise tests 275
    11.8.3 Scheffe's pairwise F-tests 278
    11.8.4 Brown–Forsyth test 280
    11.8.5 Generating one-factor data to play with 281

12 One-Factor Models with Covariates   **283**
 12.1 Introduction 283
 12.2 Exploratory Data Analysis 284
 12.3 Model Construction and Fit 286
 12.4 Diagnostics 289
    12.4.1 The constant variance assumption 289
    12.4.2 The normality assumption 291
    12.4.3 The independence assumption 292
    12.4.4 The presence and influence of outliers 292
 12.5 Pairwise Comparisons of Treatment Effects 294
    12.5.1 With a control treatment 294
    12.5.2 Without a control treatment 295
 12.6 Models with Two or More Covariates 297
 12.7 For the Curious 298
    12.7.1 Scheffe's pairwise comparisons 298
    12.7.2 The centered model 299
    12.7.3 Generating data to play with 300

13 One-Factor Models with a Blocking Variable   **301**
 13.1 Introduction 301
 13.2 Exploratory Data Analysis 303
 13.3 Model Construction and Fit 305
 13.4 Diagnostics 306
    13.4.1 The constant variance assumption 307
    13.4.2 The normality assumption 308
    13.4.3 The independence assumption 309
    13.4.4 The presence and influence of outliers 309
 13.5 Pairwise Comparisons of Treatment Effects 310
    13.5.1 With a control treatment 310
    13.5.2 Without a control treatment 311
 13.6 Tukey's Nonadditivity Test 312
 13.7 For the Curious 314
    13.7.1 Bonferroni's pairwise comparisons 314
    13.7.2 Generating data to play with 316

14 Two-Factor Models   **319**
  14.1 Introduction 319
  14.2 Exploratory Data Analysis 321
  14.3 Model Construction and Fit 322
  14.4 Diagnostics 324
  14.5 Pairwise Comparisons of Treatment Effects 325
      14.5.1 With a control treatment 326
      14.5.2 Without a control treatment 327
  14.6 What if Interaction Effects Are Significant? 328
  14.7 Data with Exactly One Observation per Cell 331
  14.8 Two-Factor Models with Covariates 331
  14.9 For the Curious: Scheffe's F-Tests 331

15 Simple Remedies for Fixed-Effects Models   **335**
  15.1 Introduction 335
  15.2 Issues with the Error Assumptions 335
  15.3 Missing Variables 336
  15.4 Issues Specific to Covariates 336
      15.4.1 Multicollinearity 336
      15.4.2 Transformations of covariates 336
      15.4.3 Blocking as an alternative to covariates 337
  15.5 For the Curious 339

Bibliography   **341**

Index   **347**

Preface

This work (referred to as Companion from here on) targets two primary audiences: Those who are familiar with the theory and applications of linear statistical models and wish to learn how to use R or supplement their abilities with R through unfamiliar ideas that might appear in this Companion; and those who are enrolled in a course on linear statistical models for which R is the computational platform to be used.

About the Content and Scope

While applications of several pre-packaged functions for complex computational procedures are demonstrated in this Companion, the focus is on programming with applications to methods used for linear regression and designed experiments with up to two fixed-effects factors, including blocking variables and covariates. The intent in compiling this Companion has been to provide as comprehensive a coverage of these topics as possible, subject to the constraint on the Companion's length.

The reader should be aware that much of the programming code presented in this Companion is at a fairly basic level and, hence, is not necessarily very elegant in style. The purpose for this is mainly pedagogical; to match instructions provided in the code as closely as possible to computational steps that might appear in a variety of texts on the subject.

Discussion on statistical theory is limited to only that which is necessary for computations; common "rules of thumb" used in interpreting graphs and computational output are provided. An effort has been made to direct the reader to resources in the literature where the scope of the Companion is exceeded, where a theoretical refresher might be useful, or where a deeper discussion may be desired. The bibliography lists a reasonable starting point for further references at a variety of levels.

About the Data

A useful skill to have for any form of modeling with the mathematical sciences is to be able to adapt methods from one application to another, analogous, application. The trick usually lies in recognizing that, while the "story" might change across applications, the essential problem remains the same. It is in this spirit that this Companion has been put together. Preparing code for applications requires guarding against quirks that invariably pop up in "real life" data. Being tied to a particular "story" can restrict the imagination of the coder and, for this reason, it is useful to prepare and test code using data that can be manipulated to possess any desired quirk.

It is a simple matter to generate data using R's random sample generators; this is how the data used for illustrations in this Companion were obtained. Some of these datasets have been given "stories," such as `lake.data`, `Retreat`, and `Salmon`, and others have not. Sample code for generating some of these datasets are presented in select chapters, and all are given in script files contained in this book's website.

About the Accompanying Website

The accompanying website for this book is located at

http://www.crcpress.com/product/isbn/9781439873656

This website contains a README file that provides brief instructions on how to use the contents, which include all data used in this Companion, all functions prepared within this Companion, and detailed script files for the code used in compiling the Companion. The script files actually contain additional code referred to, but not included in the Companion.

About the Notation

Most of the main notation used in this Companion is standard; otherwise it is defined when it first appears. Here is a summary of the main players:

Matrices and vectors: It is understood that some readers may not be familiar with these two terms; introductory descriptions are given where these terms first appear. Matrices are denoted by bold, uppercase font, for example, $\mathbf{X}$; and transposes of matrices are denoted by $\mathbf{X}'$. The entry in the ith row and j^{th} column of a matrix such as $\mathbf{X}$ is denoted by x_{ij}. Vectors are denoted by bold, lowercase font, for example, $\mathbf{x}$; and transposes of vectors are denoted by $\mathbf{x}'$. The ith entry of a vector such as $\mathbf{x}$ is denoted by x_i.

Variables and data: Variables are denoted by upper case letters; Y being reserved for the response and X, with subscripts as appropriate, for numeric explanatory variables. The exception to this rule is ε, with subscripts as appropriate, which is used to denote error terms/variables. Deviations from this notation are defined as the need arises. Data values, specified or unspecified, are denoted in the same manner as are entries of matrices and vectors.

Probabilities and quantiles: Probabilities and p-values are denoted, for example, by $P(X > x)$ and p-value, respectively. Notation such as $t(\alpha, df)$ and $F(\alpha, df_N, df_D)$ represent right-tailed quantiles.

Models: The letter p is used to denote the number of variables in a regression model, or the number of treatments in a one-factor model. Deviations from this notation are defined as the need arises. Notation specific to particular models is defined when and where needed.

R Code: All R code are given in `typewriter` font.

An effort has been made to maintain consistency in notation; however, there are some spots where clashes with earlier notation might occur. Temporary changes to notation are defined where applicable.

How to Use This Companion

Using R on "real life" data requires not only writing correct code, but also an understanding of how the code goes about producing the results it is intended to produce. Quite a bit of the code presented in this Companion can be treated as "cans of R script," which can be copied from the book's website and adapted to an application in question. However, the reader is encouraged to go through the process of typing and executing script not only to develop an intuitive feel for debugging, but also to acquire a certain degree of informed nonchalance when faced with code that refuses to cooperate. Debugging a faulty program can be infuriating, mystifying, and occasionally traumatic all in one sitting. The process of typing in basic code and encountering errors in a controlled setting helps with desensitizing oneself for later, more challenging occasions.

An effort has been made to contain all main topics within titled sections or subsections; these are listed in the table of contents. The index contains a listing of most other keywords for locating specific terms or topics that do not appear in the table of contents. If you run into unfamiliar code in an R session, use the `help` or `help.search` commands described in Section 1.3 for quick answers.

Acknowledgments

I would like to raise my hat to the R Development Core Team and the many contributors to CRAN for their efforts in providing a remarkably powerful, comprehensive, and flexible platform for statistical computing.

I would like to thank George Pearson of MacKichan Software, Inc. for his help on matters concerning Scientific Workplace, which was used to typeset this Companion in LATEX. Carrie Enge, the English and writing specialist at the Juneau Campus Learning Center for UAS, very graciously volunteered to proofread the manuscript for typographical errors, punctuation, and (strange) sentence structure. For her efforts, time, and helpful suggestions, I thank her.

Finally, I would like to thank students from my 2008, 2009, and 2010 regression and analysis of variance courses for their feedback on earlier drafts of this work. Most notable among these were Molly and Adam Zaleski, Janelle Mueller, Tyler Linderoth, Tony Gaussoin, Jennifer Stoutmore, Alex Godinez, Lorelie Smith, and Darcie Neff.

Part I

Background

Chapter 1

Getting Started

1.1	Introduction	3
1.2	Starting up R	4
1.3	Searching for Help	7
1.4	Managing Objects in the Workspace	10
1.5	Installing and Loading Packages from CRAN	12
1.6	Attaching R Objects	16
1.7	Saving Graphics Images from R	17
1.8	Viewing and Saving Session History	18
1.9	Citing R and Packages from CRAN	19
1.10	The R Script Editor	20

1.1   Introduction

R can be downloaded from the *R Project for Statistical Computing* website [56] located at

```
http://www.r-project.org/
```

Assuming this has been done, first set the stage a bit with some preliminary file management terminology and steps.

Before starting up R, reserve a space in your computer's storage area where all material associated with this R Companion might be stored. The first decision to make in performing this task is to choose a *drive* on the computer — these are typically named A, B, C, and so on; think of a drive as a "documents storage building".

Once a drive is chosen, choose a *directory* on the drive; this can be thought of as a "documents storage room" within the chosen "documents storage building". Next, create *a sub-directory* within the chosen directory; think of this as a "file cabinet" within the chosen "documents storage room." Finally, this "file cabinet" can then be made to contain as many "drawers" as desired; call these drawers *folders*. For example, R-work for preparing this companion was stored in the sub-directory named `RCompanion`, whose location is is given by

```
z:\Docs\RCompanion
```

Within this sub-directory, folders named `Chapter1`, `Chapter2`, etc. were created, and each of these contains their own *sub-folders* named `Images`, `Script`,

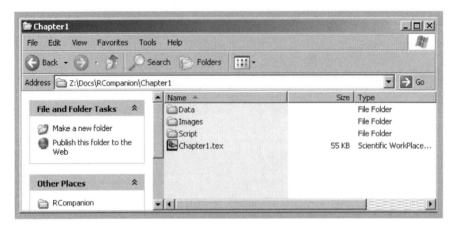

FIGURE 1.1: File and folder management; setting up folders within a working directory.

Functions, and Data; for example, see Figure 1.1. Thus, a data file, say Data01x01.txt, within the Chapter1 folder will have the *path*

z:\Docs\RCompanion\Chapter1\Data\

Think of the *path* for a file, folder, or directory as a "route" for R to follow to the location of the file, folder, or directory of interest. Setting up predetermined working and storage areas will make file saving and retrieval much more efficient once material starts piling up.

1.2 Starting up R

Figure 1.2 shows a screen shot of what is seen when R is opened and the **File** option on the menu toolbar is selected. To determine the "Start-in directory" for R on a particular computer, right-click the mouse on the **R icon** on the desktop, and then left-click the mouse on the **Properties** option. The Start-in directory can be changed by simply entering the new directory path in the icon's Properties pop-up window.

Unless instructed otherwise, R uses this Start-in directory as the working directory. The highlighted **Change dir...** option in the **File** drop down menu shown in Figure 1.2 allows the user to instruct R to change the *current working directory* within which all work in R is to be performed, and to which all files are to be saved by default. For example, by selecting the **Change dir...** option and following the directions in the **Browse For Folder** window that pops

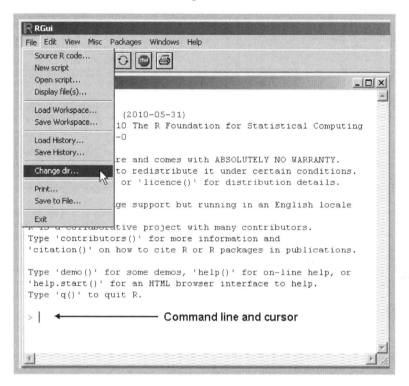

FIGURE 1.2: File management options in the R console, the cursor and command line.

up, the working directory can be changed to the Chapter1 folder mentioned previously.

This task can also be accomplished using R's *command line* feature. The command line within the *R console* (see Figure 1.2) starts at the cursor, "|", following the symbol ">". Here, a command or set of instructions for R to *execute* can be typed in, after which the **Enter** key is pressed. The *syntax* for commands in R form what has come to be known as the *R programming language*. In all that follows, the descriptors *R function, operator, syntax,* or *command* will be used interchangeably. For example, the working directory can also be set by executing the command

```
> setwd("Z:/Docs/RCompanion/Chapter1/")
```

It is important to remember that *R is case sensitive*; lower and upper case letters are treated as different letters so commands have to be entered exactly as they are defined. The first few sections of this chapter focus on working in the command line mode. The last section introduces the *R Script Editor* within which *R programs*, or logical sequences of code using R syntax, are written and then executed. R uses these programs to perform computations or produce graphics, which are displayed in *R Graphics Device* windows.

Observe that when defining the path of a file or folder in R, a forward-slash, "/" is used as opposed to a backward slash,"\". This syntax works for Windows, MacIntosh, and UNIX operating systems. In Windows (as opposed to other operating systems) one may use two backward-slashes, "\\", when defining file or folder paths. It should be noted that the syntax presented in this companion is for R in the Windows environment; however, the majority of the given syntax also works in the MacIntosh environment. Mac users will find that for the few differences that do exist, the help resources available through the MacIntosh version of R will suffice in making appropriate corrections.

Commands executed in R often create *objects*, which are stored in what is called the *workspace*. The R workspace can be thought of as that part of the active computer memory that R uses. Objects, in R, can be thought of as named locations in the workspace that may contain a variety of structures including, for example, variable lists, tables, and functions, to mention a few. The current workspace can be saved at any time during a session using the drop down menu option **File→Save Workspace** (see Figure 1.2). A prompt, in a **Save image in** window, for a file name appears at which time a sub-folder can also be selected. The file name should have the extension "**.RData**". The named workspace file is then saved in the chosen folder within the current work directory. This task can also be performed at the command line level by executing, for example,

```
> save.image("Z:/Docs/RCompanion/Chapter1/scratch.RData")
```

Note that specifying the path is necessary in the `save.image` command. Once a workspace has been saved, it may be recalled at any time, along with all the objects created within it. *Loading a saved workspace* can be achieved using the drop down menu option **File→Load Workspace...** (see Figure 1.2), and then selecting the appropriate folder within the working directory and the desired workspace file; R goes to the current working directory unless otherwise directed. This task can also be accomplished in the command line environment by executing, for example,

```
> load("Z:/Docs/RCompanion/Chapter1/scratch.RData")
```

To *exit R*, use the drop down menu option **File→Exit** (see Figure 1.2) or the command `q()` in the command line environment. In either case, a pop-up prompt will appear asking whether the current workspace should be saved. If **Yes** is selected and the workspace is saved in the file named simply ".RData" in the *Start-in directory*, the directory in which R has been instructed to start up, then when R is opened at a later time, the saved workspace is automatically uploaded, this being indicated by the following message on the console:

```
[Previously saved workspace restored]
```

If the workspace is saved in any other directory, the previously saved workspace will not be restored. If **No** is selected, the workspace is not saved anywhere.

After long sessions in R, the console will contain a long list of command executions and output lines. If this is a bother, it is a simple matter to *clear the console* to tidy things up using the toolbar option **Edit→Clear console** or the key sequence **Ctrl + L**; keep the **Ctrl** key pressed and then press the **L** key on the keyboard. Be aware that clearing the console does not remove anything from the workspace.

The **Windows** drop down menu on the R console allows users to toggle between the *R Console*, *Script Editor*, and *Graphics Device* windows.

1.3 Searching for Help

Probably the most helpful set of options in the console menu bar is in the **Help** drop down menu (see Figure 1.3). Here is a very brief outline of the help menu options: Selecting **Console** will display quick instructions on navigating the R console in the command line environment. The **FAQ of R** and **FAQ on R for Windows** (or MacIntosh) are fairly self-explanatory; selection of either of these items will result in windows with commonly asked questions for R in general and also for Windows and Mac users. These options might serve as useful first stops in seeking answers to questions. **Manuals (In PDF)** and all of the remaining items are also fairly self-explanatory. The **Search help...** option is easiest to use if a collection of keywords that roughly describe what is wanted is known. Results from this option might (or might not) help to zero in on specific help resources. Another good source is the **Html help** option, which opens an online resource page. It is also possible to perform very effective searches in the command line environment.

Syntax and strategies for searching the **R Documentation** resources for help in the command line environment are (for the most part) fairly intuitive. As an example, suppose information on basic arithmetic operations in R is desired. Then, an appropriate first step would be to use the `help.search`. For example, one might expect the command

```
> help.search("arithmetic")
```

to get results. In this case, R opens an **R Information** window with the following information

```
Help files with alias or concept or title matching
'arithmetic' using fuzzy matching:
base::+ Arithmetic Operators
```

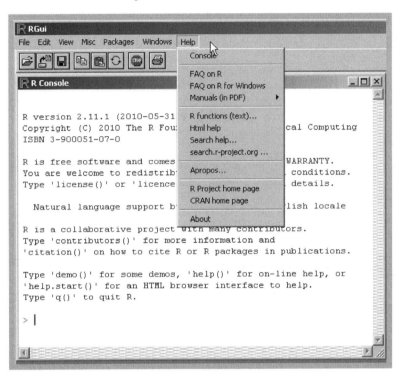

FIGURE 1.3: The R Help menu for use in finding help on various topics such as functions, packages, and general FAQs.

```
base::mean Arithmetic Mean

stats::weighted.mean Weighted Arithmetic Mean

Type '?PKG::FOO' to inspect entry 'PKG::FOO TITLE'.
```

Of the choices given, the most likely candidate is `Arithmetic Operators`. The `base::+` notation indicates that `Arithmetic Operators (TITLE)` are located in package (`PKG`) `base` [56] and the R Documentation file on the operator(s) is named "+" (`FOO`). Executing the command

```
> help("+",package=base)
```

opens an **R Documentation** page on Arithmetic Operators. Using "−", "∗", "/" or "^" in place of "+" will have the same effect as long as the symbol is enclosed in quotes.

There are times, however, when typing in and executing the `help` command will go clearly wrong, typically in one of two ways: A message of the form

```
> help(+,package=base)
```

```
Error: unexpected ',' in "help(+,"
```

suggests R is misinterpreting the argument +; R wants to perform an addition and does not like the comma that follows. Such messages suggest an error in the argument entry. On the other hand, a message of the form

```
> help(x,package=base)
No documentation for 'x' in specified packages and
libraries:
you could try '??x'
```

indicates an unknown or incorrect R function name was used — look for typographical errors.

The help.search command also kicks out messages. For example,

```
> help.search("brown-forsyth test")
No help files found with alias or concept or title matching
'brown-forsyth test' using fuzzy matching.
```

This means the keywords used are not helpful. Out of curiosity, try

```
> help.search("fuzzy matching")
```

It is nice to know that the help.search command is sometimes helpful in interpreting messages from R.

One may wish to explore the contents of a particular package. As an example, by executing

```
> help(package=base)
```

a **Documentation for package base** window opens, giving a brief description of the package and its contents. Package base is one of a few default packages that are automatically loaded when R is started up. There are times when this command will not work. The reasons for this are typically an incorrectly typed name or, sometimes, the package of interest is not installed or loaded (to be discussed shortly) on the computer. In such cases, a command of the form

```
> library(help=base)
```

will probably work. A similar command, using any package name, will bring up the R Documentation page on the identifed package (if it exists), whether or not it is installed or currently loaded for direct access; for example,

```
> help(base,package=base)
```

opens the R Documentation page for package `base`. By clicking on the `Index` at the bottom of this page, the whole list of functions and operators contained in package `base`, along with links to their own documentation pages appears. Failed searches with this last help command are usually a result of entering an incorrect package name. It appears that all of the above `help` commands work whether or not the package name is enclosed in double quotes; however, in case an error message occurs, an option would be to try executing the command with the package name in double quotes, for example,

```
> help(base,package="base")
```

In summary, help on any operator, function, or package in R can be obtained directly if specific names are known, or indirectly through keyword searches. To find out in which package a function, say `anova`, is contained, execute

```
> help.search("anova")
```

In the R Information window, among a large number of fuzzy matches, it will be found that

```
stats::anova           Anova Tables
```

Thus, the function `anova` is contained in package `stats` [56].

1.4 Managing Objects in the Workspace

If, during an R session, a new package or workspace is loaded, R will not delete objects present in the current working memory. The objects contained in the new package or workspace are simply added to the list of objects in memory. After a while it may be that too many objects are stored in the computer's memory and this may lead to problems when two or more objects compete or conflict in some manner. Examples of how such conflicts manifest themselves appear periodically in later chapters. Careful bookkeeping and workspace management thus becomes a necessity to ensure R runs smoothly.

To help illustrate the use of the various bookeeping commands, begin by first executing the following commands

```
> x<-1:25; y<-2*x+3; z<-3*x-2*y
> pairs<-data.frame("x"=x,"y"=y)
> triples<-data.frame("x"=x,"y"=x,"z"=x)
```

Observe that a sequence of distinct commands can be executed on the same command line if they are separated by a semicolon, ";".

The function `ls` and the function `objects`, both from package `base`, can be used to list all objects contained in the workspace. For example, execution of the following first line results in the output given in the second line.

```
> ls()
[1] "pairs" "triples" "x" "y" "z"
```

The [1] indicates the output order of the first entry in the corresponding output line. Executing the command `objects()` produces the same output.

Specific objects contained in the workspace can be removed; for example,

```
> rm("x","y","z")
```

removes the objects x, y, and z from the workspace — check by executing `ls()`. Note that an alternative form of the previous command is

```
> rm(list=c("x","y","z"))
```

The *combine function*, "c", combines the names "x", "y" and "z" into a single list of three names. More on the combine function will be discussed in Chapter 3; for the present it is important to remember that, when used in the `rm` function, object names within the function `c` must always be enclosed in quotes; otherwise, R will first combine the three lists of numbers into a single list, which has no name and, hence, cannot be removed.

There are times when it may be advantageous to save an object, such as a data file, for later recall. The object `triples` might qualify as one such object, then

```
> dump("triples",
        "Z:/Docs/RCompanion/Chapter1/Data/triples.R")
```

saves the object `triples`, which must be enclosed in quotes, as the file `triples.R` in the Data folder as indicated by the given path.[1] Such files must always have the extension ".R". Other options for the `dump` function include, for example,

```
> dump(c("pairs","triples"),
        "Z:/Docs/RCompanion/Chapter1/Data/pairsntriples.R")
> dump(list=ls(all=TRUE),
        "Z:/Docs/RCompanion/Chapter1/Data/alldata.R")
```

---

[1]This command, and the two `dump` commands following should be interpreted, and entered into the console, as a single line.

The first of these two commands stores the objects `pairs` and `triples` in the file `pairsntriples.R`, and the second stores all objects in the workspace in the file `alldata.R`.

Any of these saved files can be recalled whenever desired. For example, the command

> `source("Z:/Docs/RCompanion/Chapter1/Data/triples.R")`

loads the object `triples` in the workspace. This recall can also be accomplished using the sequence **File→Source R code...** (see Figure 1.2).

To remove all objects in the workspace, execute

> `rm(list=ls(all=TRUE))`

Be careful not to execute this command unless it is known that none of the objects in the workspace will be needed later in a session.

1.5   Installing and Loading Packages from CRAN

The *Comprehensive R Archive Network (CRAN)* is located on the R-Project homepage, and contains resources in the form of packages developed by users of R from all over the world. These packages provide R users with an incredible resource for functions that have applications ranging from very broad to extremely specialized. There are two descriptors that show up in the following discussion: *loaded packages* and *installed packages*.

A package that is installed on the computer and then loaded into the working memory is one whose contents are readily available for use. Package `base` is an example of one such package. As mentioned earlier, some packages are loaded by default when R is started up, and it is useful to know which these are. A quick way to find out which packages are currently loaded in the computer memory is to execute

> `search()`

If no new packages are loaded, and depending on the width of the R Console window, the subsequent output may have the appearance

```
[1] ".GlobalEnv"      "package:stats"    "package:graphics"
[4] "package:grDevices" "package:utils"   "package:datasets"
[7] "package:methods"  "Autoloads"        "package:base"
```

As a side note, observe that the second line begins with [4], indicating the second line begins with the fourth output string. Similarly, the third line begins with [7], the seventh output string.

A package that is installed on the computer, but not loaded, is one that can be called up for use at any time and then put back on the "shelf" when it is no longer needed. The standard R build appears to come installed with packages that contain functions to perform what might be considered common tasks. There are a couple of ways in which the packages installed on a computer (as opposed to loaded in the working memory) can be listed. One is by executing

```
> library()
```

This opens an **R packages available** window with a list of all packages available for direct *loading* into the working memory. Another option is to execute

```
> installed.packages()
```

This outputs the same list, but in the console and with a lot of technical details.

When confronted with the task of using R for computations for which no function appears to be available in any of the packages installed in the R library, there are two options: One would be to develop a program or function to perform the task; the other, often more efficient, option is to perform a search for resources using methods mentioned earlier, or the Wiki feature on the R Project homepage, or even Google. Very often, a search will lead to an appropriate function written by someone and stored in a package available for downloading from CRAN. If the target package is not present in *your* R library, you may install it from CRAN.

As an example of downloading packages from CRAN, consider installing the packages: car [26]; ellipse [52]; faraway [23]; HH [39]; leaps [47]; lmtest [71]; pls [70]; and quantreg [44]. Functions from these packages will find use in later chapters.

Package Installation Steps: Refer to Figure 1.4 and follow the steps given below — Internet access is needed to do this.

1. Click on the **Packages** button and select **Install packages...**

2. A prompt will appear to select a CRAN mirror from the **CRAN mirror** window. Choose a mirror, say **USA (WA)**, and then click on **OK**.

3. A list of packages will appear in the **Packages** window. Keep the **Ctrl** key pressed and, in the **Packages** window, select the packages listed above by left-clicking your mouse on each package — the result should be that all (and only) the desired packages will be highlighted. Then click on **OK**.

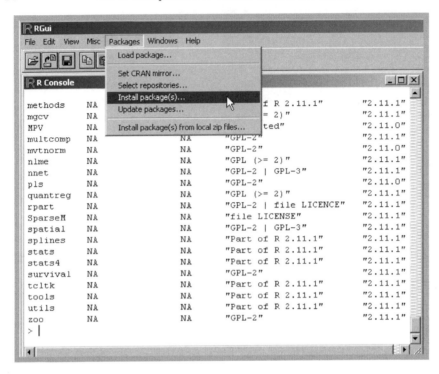

FIGURE 1.4: Using `Packages` menu options to identify and install packages from CRAN.

4. To check that the download was successful, enter `library()`. The output for this command will include all the newly installed packages along with the original list.

To use the newly installed packages, or any others that are installed but not loaded, the desired package can be loaded by executing, for example,

> `library(car)`

or

> `require(faraway)`

While the two commands perform equivalent tasks, the function `require` has a safety feature for avoiding error messages that makes it suited for use inside other functions. See `help(library,package=base)` for details. There is one thing that becomes evident if package `car` and then package `faraway` is loaded. On execution of the second command, the output

```
Loading required package: faraway
Attaching package: 'faraway'
The following object(s) are masked from 'package:car':
    logit, vif
```

seems to suggest a clash has occurred. The above message states that packages `car` and `faraway` both contain objects (functions) named `logit` and `vif`. While this scenario may not always create problems, there will be occasions when the use of such masked functions will result in incorrect results, particularly if the commonly named functions are not identical in all respects.

The safest approach to using packages that are not in the default list is to load only one at a time, then be sure to remove each one once it is no longer needed and *before* any other nonstandard package is loaded. For example, once package `car` is no longer needed, it can be removed from the working memory using

```
> detach(package:car)
```

Similarly, package `faraway` may be removed; it appears that packages can only be detached one at a time. If package `car` had not been loaded prior to package `faraway`, the above "masked" message would not have been printed on the console. There is also the possibility of loading the same package several times. While this does not appear to create problems, it can be avoided by making use of the `search` command as described earlier.

Be aware that some packages depend on other packages. If such a package is loaded, R automatically loads all other packages needed by the desired package. As an example,

```
> library(HH)
Loading required package: lattice
Loading required package: grid
Loading required package: multcomp
Loading required package: mvtnorm
Loading required package: survival
Loading required package: splines
Loading required package: leaps
```

Executing the command `search()` will show that in addition to the default packages and package HH, each of the above packages is also loaded.

When detaching all of these surplus packages after they are no longer needed, it is important to do the detaching in the same order as they were loaded starting with the featured package, HH. If this is not done, R will refuse to detach a package if it is needed by another currently loaded package. For package HH, the task is accomplished by executing the following commands:

```
> detach(package:HH); detach(package:lattice)
> detach(package:grid); detach(package:multcomp)
> detach(package:mvtnorm); detach(package:survival)
> detach(package:splines); detach(package:leaps)
```

does the task, except for some unnerving warning messages. In this particular
case, R lets the user know that it could not find certain objects or functions
to clean up. This is probably because each of the objects or functions belong
to, or are contained in, another package. Here, there are no worries.

There are occasions when, on loading a package, a message warns the user
of the version or date of the package being loaded. This message simply alerts
the user to be aware that the package may be outdated. For the most part,
functions contained in such packages work fine because of the consistency in
the base syntax for R.

1.6   Attaching R Objects

For the following discussion, first execute

```
> rm(list=ls(all=T))
```

(note the shorthand version for TRUE), and then recall the earlier saved objects,
`pairs` and `triples`, by executing, for example,

```
> source("Z:/Docs/RCompanion/Chapter1/Data/
                              pairsntriples.R")
```

On executing `ls()`, it will be seen that the objects `pairs` and `triples` are
listed as the only objects contained in the workspace.

Both of these objects contain *sub-objects*, for example, by executing

```
> names(pairs)
```

it is seen that the object `pairs` contains two sub-objects, x and y. Similarly,
`triples` contains the sub-objects x, y, and z. These sub-objects cannot be
directly accessed, for example, by entering simply x; the result will be an
error message of the form

```
Error: object 'x' not found
```

However, by first executing

```
> attach(pairs)
```

and then entering x, the result might look like

```
> x
[1]  1  2  3  4  5  6  7  8  9  10  11  12  13  14  15  16  17  18  19  20  21
[22] 22 23 24 25
```

It will also be noticed that the command

```
> search()
```

reveals `pairs` in the list of loaded packages (and now objects). While there may be times when direct access to sub-objects is desireable for writing more efficient or tidier code, as with packages, caution must be exercised when attaching objects. For example, consider attaching the object `triples` to the workspace as well. Then,

```
The following object(s) are masked from 'pairs':
x, y
```

If an object remains attached after its purpose has been served, one of two things can happen: The contents of the named sub-objects might be inadvertantly altered; or the named sub-objects may interfere with identically named sub-objects of another, later, attached object.

If at all possible, avoid the use of the `attach` command. If there is no alternative, remember to *restrict* access to sub-objects within their parent object once the task at hand is accomplished. This is done by executing, for example,

```
> detach(pairs); detach(triples)
```

It appears that, as with detaching loaded packages, attached objects need to be detached one at a time.

Another point to note here is that an object can be attached more than once; however, R will produce a mask-alert if this is done. See `help(detach,package=base)` for further details and other tips.

1.7   Saving Graphics Images from R

To illustrate the process, execute the commands

```
> win.graph(width=3.5,height=3); par(ps=10,cex=.75)
> plot(pairs)
```

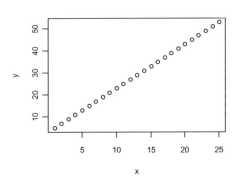

FIGURE 1.5: A graphics image created in the R graphics device window, and then inserted in a LaTeX document.

The result will be a scatterplot of y against x in the **R Graphics Device** window (see Figure 1.5).

Chapter 4 covers R graphics capabilities in considerable detail; for the present, all that is of interest is how to export the image for insertion into a file produced by word processing programs such as MS Word or LaTeX.

If the mouse is right-clicked on the Graphics Device window, a popup window indicates there are two copy options (as a metafile and as a bitmap), two save options (as a metafile and as a postscript file), and a print option. The copy options permit the familiar *copy-and-paste* process and the print option sends the image to a printer.

Consider saving the image in postscript format (having an ".eps" extension). To do this, right-click on the image, select **Save as postscript...**, then save the image as `Figure01x05.eps` in the `Image` sub-folder of the working directory `Chapter1`. This image can then be imported into the word processing file as appropriate.

1.8   Viewing and Saving Session History

It is sometimes helpful to recall command sequences entered in an R session. Suppose the last 25 lines of code in this session are of interest, then execute

```
> history(max.show=25)
```

The argument `max.show` instructs R to display the last 25 command lines, the twenty-fifth line being the above command. An **R History** window will pop up showing the last 25 lines of commands. If `max.show=Inf` is used, then R will display up to a default of 512 lines of commands.

Another useful option is to save the history of a full R session. This can be done, for example, by executing

```
> savehistory("Z:/Docs/RCompanion/Chapter1/
                                     Chapter1history.txt")
```

which creates a text file of all commands executed in a session. The text file can then be viewed in any text editor, such as Notepad and even MS Word.

This task can also be accomplished (see Figure 1.2) using the menu sequence **File→Save History**... to save to a specific file (with extension ".Rhistory" or ".txt"). This file can be loaded back into the workspace memory using **File→Load History....** and the current session history can then be attached to the previous session history by another save. The contents of a session history file can be viewed using a text editor or the R *script editor*, which is introduced in the last section of this chapter.

The command `savehistory()` will save the full session to the default file ".RHistory" in the current working directory. As with the default saved workspace image file ".RData", if the file ".RHistory" is saved in the Start-in directory for R, then the previous session history is automatically restored when R is started up and the current session history can be attached to the previous session history by executing the `savehistory` command.

1.9 Citing R and Packages from CRAN

While R and packages in CRAN are in the public domain, the efforts of all contributers to this powerful statistical computing resource should be acknowledged appropriately in any publication that makes use of R for data analysis. Information on how to cite R in publications can be obtained by executing the command `citation()`. For citation details on specific packages from CRAN execute a command of the form

```
> citation("car")
John Fox. I am grateful to Douglas Bates, David Firth,
Michael Friendly, Gregor Gorjanc, Spencer Graves, Richard
Heiberger, Georges Monette, Henric Nilsson, Derek Ogle,
Brian Ripley, Sanford Weisberg, and Achim Zeileis for
various suggestions and contributions. (2009).
car: Companion to Applied Regression. R package version
1.2-16.
http://CRAN.R-project.org/package=car
```

1.10   The R Script Editor

All R commands for this and later chapters were entered into the *R script editor*, and then executed. This is a much more convenient setting in which to work, so a brief illustration is given here.

Among the menu options listed in Figure 1.2 is **New script**. If this option is chosen, an **Untitled - R Editor** window opens. This can then be saved as a named file, for example, `Chapter1Script.R`, in a `Script` folder within the current working directory as follows: With the R editor window active, select **File→Save as ...** as shown in Figure 1.6. The **Save script as** window, shown in Figure 1.7, permits selecting the desired folder and assigning the file name. Make sure the file name contains the extension ".R". On entering code, the file may be saved periodically, and when a session ends the file can be saved and closed for later recall.

Type commands in a script file as in any text editor. Executing commands is also straightforward. Here is an illustration using the `plot` command from earlier (see Figure 1.8). Use the mouse (left-press) to highlight the sequence of commands that are to be run, then right-click the mouse and select the **Run line or selection** option by left-clicking the mouse. The console will show

```
> win.graph(width=3.5,height=3); par(ps=10,cex=.75)
> plot(pairs)
```

and the R Graphics window will open with the earlier plot.

For future reference, there will be occassions when a command will be too long to fit in a single line. When such commands are entered into the script editor, it is best to indent the continued command on subsequent lines to add clarity. An important point is to make sure that R expects the command to be continued. Here are a couple of illustrations from a later chapter. The first illustration shows a correct extension of a long command to a second line (see Figure 1.9).

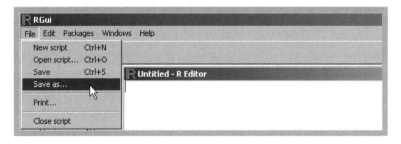

FIGURE 1.6: Opening, naming, and saving a script file using script editor menu options for later use or editing.

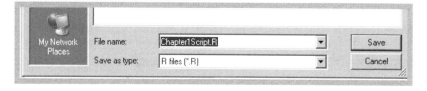

FIGURE 1.7: The `Save script as...` window. Remember to place a `.R` at the end of the file name.

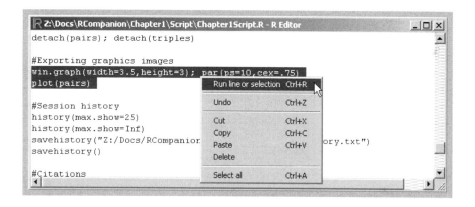

FIGURE 1.8: Executing single and multi-line commands within the script editor.

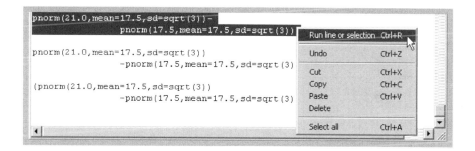

FIGURE 1.9: Entering and structuring multi-line commands appropriately in the script editor.

On executing the highlighted lines in Figure 1.9, the console shows

```
> pnorm(21.0,mean=17.5,sd=sqrt(3))-
+                   pnorm(17.5,mean=17.5,sd=sqrt(3))
[1] 0.4783459
```

The "+" on the second line of the command indicates the command from the first line continues to give the output following the [1].

On the other hand, if the following code were to be put into the script editor and then run

```
pnorm(21.0,mean=17.5,sd=sqrt(3))
                    -pnorm(17.5,mean=17.5,sd=sqrt(3))
```

the console shows

```
> pnorm(21.0,mean=17.5,sd=sqrt(3))
[1] 0.978346
>                   -pnorm(17.5,mean=17.5,sd=sqrt(3))
[1] -0.5
```

Here, moving the "-" to the second line makes the first line a complete command, so it is executed before moving to the second line. The same is the case with the second line. An alternative way to correctly write the above is to enclose the whole command within parentheses. For example, running the command

```
(pnorm(21.0,mean=17.5,sd=sqrt(3))
                    -pnorm(17.5,mean=17.5,sd=sqrt(3)))
```

produces

```
> (pnorm(21.0,mean=17.5,sd=sqrt(3))
+                   -pnorm(17.5,mean=17.5,sd=sqrt(3)))
[1] 0.4783459
```

So, when extending a command to two or more lines, it is important to make sure that R is left hanging until the end of the command is reached.

Chapter 2

Working with Numbers

2.1   Introduction 23
2.2   Elementary Operators and Functions 23
    2.2.1   Elementary arithmetic operations 24
    2.2.2   Common mathematical functions 24
    2.2.3   Other miscellaneous operators and functions 25
    2.2.4   Some cautions on computations 26
2.3   Sequences of Numbers 27
    2.3.1   Arithmetic sequences 27
    2.3.2   Customized sequences 28
    2.3.3   Summations of sequences 28
2.4   Common Probability Distributions 29
    2.4.1   Normal distributions 29
    2.4.2   t-distributions 30
    2.4.3   F-distributions 31
2.5   User Defined Functions 32

2.1   Introduction

This chapter can be viewed as an outline of doing arithmetic and basic statistical computations in R. The tasks illustrated involve the fundamentals needed to perform more complex computations in later chapters.

Note that from here on the symbol ">" at the beginning of executed command lines, and the bracketed number preceding each output line, for example [1], will be left out except where the distinction between a command line and an output line is not clear. Also, all code for this and later chapters are entered and run in the R script editor.

2.2   Elementary Operators and Functions

The two broad classifications of numbers used in R for this Companion are *integers*, such as $\{\ldots -3, -2, -1, 0, 1, 2, 3, \ldots\}$ and *real numbers*, any number that is not an integer. Recall, *rational numbers* are those that have terminating *or* repeating decimal representations, while *irrational numbers* are those that have non-terminating *and* non-repeating decimal representations. In R, real numbers are output in *decimal form* (for example, when 1/3 is entered the output is 0.3333333), or in *scientific notation* (for example, when 25^26 is entered the output is 2.220446e+36).

It is important to remember that, for the most part, R works with *rational approximations* of real numbers; for example, for $\sqrt{2}$ a rational approximation up to 11 decimal places appears to be used and the same for e and π. While computations *within* R (probably) use approximations up to a certain high level of precision, the output of computed results typically seem to have at most seven decimal places, with or without scientific notation. This level of accuracy is usually more than adequate in applications.

2.2.1   Elementary arithmetic operations

The usual *arithmetic operations* of addition, subtraction, multiplication, division, and exponentiation on numbers apply in R. Moreover, as long as the operations to be performed are mathematically legal, entered commands will work as should be expected. See any standard college algebra or pre-calculus text if a refresher on algebra or common mathematical functions is needed.

Let a and b be real numbers. Then, subject to appropriate restrictions,

Operation	Syntax	Example	Output
$a + b$	a+b	5+7	12
$a - b$	a-b	5-7	-2
$a \times b$	a*b	5*7	35
$a/b,\ b \neq 0$	a/b	5/7	0.7142857
$b^a,\ b > 0$	b^a	5^7	78125

Note that for b^a if $0 < a < 1$ and $b < 0$, R will output NaN (Not a Number). Also for b^a, if $b = 0$ and $a < 0$, R will output Inf.

2.2.2   Common mathematical functions

Given below is a list of some common functions, with examples, that may be useful in later chapters. All of these functions are contained in package base. As with any mathematical function, attention should be paid to domains of definition when using these functions.

Function	Syntax	Example	Output
$\sin^{-1}(x)$	asin(x)	asin(sqrt(2)/2)	0.7853982
$\ln(x)$	log(x)	log(5)	1.609438
e^x	exp(x)	exp(3.2)	24.53253
$|a|$	abs(a)	abs(-5)	5
$\sqrt{a}$	sqrt(a)	sqrt(5)	2.236068
$\sqrt[n]{a}$	a^(1/n)	5^(1/7)	1.258499

Note that if $a < 0$, R will output NaN for any root.

2.2.3 Other miscellaneous operators and functions

Estimates of e and π can be obtained as shown below:

Constant/Function	Syntax	Example	Output
e	exp(n)	exp(1)	2.718282
π	pi	2*pi	6.283185

Let a, b, and n be positive integers. Then, the notation $a \bmod b$ represents the remainder obtained when a is divided by b and the notation $a \operatorname{div} b$ represents the integer quotient obtained when a is divided by b. These computations are illustrated below.

Next, let n be a positive integer, then the factorial notation, $n!$, represents the product $n \times (n-1) \times (n-2) \times \cdots \times 2 \times 1$, and the combination notation nC_k represents the number of ways k objects can be chosen from n objects. These computations are also illustrated below.

Operator/Function	Syntax	Example	Output
$a \bmod b$	a%%b	12%%5	2
$a \operatorname{div} b$	a%/%b	123%/%17	7
$n!$	factorial(n)	factorial(7)	5040
nC_k	choose(n,k)	choose(7,2)	21

Sometimes it may be necessary to input numbers in scientific notation for computations in R. Let $0 < a < 10$ and let n be any non-zero integer. Here is how this is done in R:

To Input	Syntax	Example	Output
$a \times 10^n$	aen	(3.21e-5)*(1.23e7)	394.83

It appears that very large or very small numbers are stored and outputted in scientific notation.

There are times when the results of a computation may need to be reduced in some fashion. Familiar reductions include:

Function/Syntax	Example	Output
round(x, n)	round(pi,5)	3.14159
ceiling(x)	ceiling(pi)	4
floor(x)	floor(pi)	3
trunc(x)	trunc(pi)	3
signif(x, n)	signif(3.567898272e+5,2)	360000

The default value for n in the function round is 0, and for signif the default value for n is 6. An equivalent to the function trunc(x) is the function as.integer(x).

Sometimes there is a need to determine the sign of a number, or determine the classification type (such as *integer, real, numeric, character*, etc.) of a data value. R functions available for this are:

Function/Syntax	Example	Output
sign(*x*)	sign(-3); sign(0); sign(3)	-1, 0, 1
is.integer(*x*)	is.integer(pi)	FALSE
is.real(*x*)	is.real(pi)	TRUE
mode(*x*)	mode(FALSE)	"logical"

Other data classification type testing functions are available, all of which have the appearance is.*datatype*, where *datatype* is the classification type of interest.

2.2.4   Some cautions on computations

It is important to keep in mind that R uses approximations in the majority of its computations, so having a conceptual understanding of what is entered is essential to ensure meaningful results. Consider the following:

```
> tan(pi/2)
[1] 1.633178e+16
> log(0)
[1] -Inf
```

However, $\tan(\pi/2)$ and $\ln(0)$ are undefined. Similarly,

```
> asin(1.1)
Warning message: NaNs produced
```

Note that $\sin^{-1}(1.1)$ does not exist, hence the NaNs. The term NaN translates to "Not a Number". Another case is

```
> (-27)^(1/3)
[1] NaN
```

but $\sqrt[3]{-27} = -3$. So, sometimes a false NaN is outputted.

Observe that sometimes R will detect a problem and return an error or a warning message; other times it will provide an answer that may not be mathematically correct.

Directions and translations for most issues can be searched for in R, but it will take getting used to figuring out search strategies. Explore by using a variety of keywords; some of the results you get may be quite far from what you want or expect. Remember, if a search turns out unsuccessful within a particular package, it may be that the item you are looking for is not in the package and the help.search command should be used to cast a wider net.

2.3   Sequences of Numbers

A sequence of n numbers can be represented by the notation $x_1, x_2, \ldots, x_n$ where x_i represents the ith number in the sequence. A variety of classes of sequences will find use in later chapters; here are some examples.

2.3.1   Arithmetic sequences

A sequence of n consecutive integers, starting with the integer a and ending with the integer b, has the appearance

$$a, a+1, a+2, \ldots, b$$

where $b = a + (n - 1)$. In R, any such sequence can be obtained as follows:

Operator/Syntax	Example	Output
a:b	0:100	0 1 2 3 ... 98 99 100

Note that in using the *colon operator*, ":", the numbers a and b need not be integers, and it is not necessary for $a < b$ to be the case in order for a sequence to be generated. Try using a variety of values for a and b, some for which $a < b$ and some for which $a > b$.

Consider an increasing sequence of n numbers, starting with the number a and ending with the number b. Now, suppose all consecutive numbers in this sequence differ by a *fixed* value of d, for example $x_5 - x_4 = d$. Then, the terms of the sequence have the appearance

$$a, a+d, a+2d, \ldots, b$$

where $b = a + (n - 1)d$. Notice that the previous sequences are special cases of such sequences with $d = 1$. Such sequences are called *arithmetic sequences*, and these may be generated as follows:

Function/Syntax	Example	Output
seq(a,b,d)	seq(0,100,2)	2 4 6 8 ... 96 98 100

If $a < b$, then d needs to be positive, and an increasing sequence is generated. If $a > b$, then d needs to be negative, and a decreasing sequence is generated. It is helpful, but not necessary, for the difference of a and b to be an integer multiple of d, in which case the number of terms generated by seq will be $n = |b - a|/d + 1$ and the last term in the sequence will exactly equal b.

Try a variety of values for a, b, and d, and see if you can force some error messages.

2.3.2 Customized sequences

Notice that both sequences described above are in one-to-one correspondence with the set of natural numbers. It is possible to obtain more general sequences, also in one-to-one correspondence with the natural numbers. Consider generating a new sequence from the integer sequence `0:n` using a function f where the sequence generated has the appearance

$$f(0), f(1), \ldots, f(n).$$

This can be done simply by using a command of the form `f(0:n)`, as long as the function `f` is defined in the workspace. Here is an example:

Consider generating a *geometric sequence* of the form

$$a, a\,r, a\,r^2, \ldots, a\,r^n.$$

If i is an integer satisfying $0 \le i \le n$, each term of the sequence has the functional appearance

$$f(i) = a\,r^i,$$

and the first $n + 1$ terms of the geometric series starting at a and having *common ratio* r can be obtained.

```
> 2*2^(0:9)

[1] 2 4 8 16 32 64 128 256 512 1024
```

Note that the function `f` needs to be known; in this case `f` is `2*2^i`, where values for `i` are obtained from `0:9`.

See also the R documentation pages for the function `sample` which can be used to extract a random sequence (sample) from a larger set of data.

2.3.3 Summations of sequences

The sum of any sequence, that is, $x_1 + x_2 + \cdots + x_n$, may be obtained using the `sum` function. For the geometric sequence above,

```
> sum(2*2^(0:9))

[1] 2046
```

The logical parameter `na.rm` can also be passed into the function `sum`. This can be assigned the values `TRUE` (`T`) or `FALSE` (`F`) to indicate whether missing values should be removed. If `na.rm=FALSE`, an `NA` (not available) value in any of the arguments will cause an overall value of `NA` to be returned; otherwise, if `na.rm=TRUE`, `NA` values are ignored and R returns the sum of all available values. Not including this parameter in the command line will cause R to default to `na.rm=FALSE`.

2.4   Common Probability Distributions

The R documentation page for package `stats` can be opened using `help(stats,package=stats)`. The Index link at the bottom of the page then leads to the full list of functions available in this package. You can also use the keyword "distribution" in a search for all probability distribution functions available in R. Functions for three of the distributions available in R are demonstrated here.

2.4.1   Normal distributions

Let X be a normally distributed random variable with mean μ and standard deviation σ, and denote the density function of X by $d(x)$. Note that since X is a continuous random variable, $d(x)$ does not represent a probability.

Function/Syntax	Computes
`dnorm(x,mean=`μ`,sd=`σ`)`	$d(x)$
`pnorm(x,mean=`μ`,sd=`σ`,lower.tail=T)`	$P(X \leq x)$
`qnorm(p,mean=`μ`,sd=`σ`,lower.tail=T)`	x for which $P(X \leq x) = p$
`rnorm(n,mean=`μ`,sd=`σ`)`	Random sample of size n

If the argument `lower.tail` is left out of `pnorm` or `qnorm`, the default value of `lower.tail=T` is assigned. If `lower.tail=F` is used in these functions, then R computes probabilities and quantiles associated with the right tail of the corresponding distribution.

If, in any of the above commands, the parameters `mean` or `sd` are not specified, R assumes the default values of 0 or 1, respectively. Here are some illustrations.

Suppose X is normally distributed with parameters $\mu = 17.5$ and $\sigma = \sqrt{3}$. Then, to find $P(X \leq 17.5)$, use

```
> pnorm(17.5,mean=17.5,sd=sqrt(3))
[1] 0.5
```

To find $P(X > 21.0)$, use

```
> pnorm(21.0,mean=17.5,sd=sqrt(3),lower.tail=F)
[1] 0.02165407
```

To find $P(17.5 < X < 21)$, use the following. Recall that if a command line is long, it can be continued on a second line.

```
> pnorm(21.0,mean=17.5,sd=sqrt(3))-
+           pnorm(17.5,mean=17.5,sd=sqrt(3))
[1] 0.4783459
```

Note that when a multi-line command is executed, the continuation to a new line is indicated by a leading "+" in the new command line.

To find x for which $P(X \leq x) = 0.05$, use

```
> qnorm(0.05,mean=17.5,sd=sqrt(3))
[1] 14.65103
```

and to find x for which $P(X > x) = 0.05$, use[1]

```
> qnorm(0.05,mean=17.5,sd=sqrt(3),lower.tail=F)
[1] 20.34897
```

Finally, to obtain a random sample of size 5, use

```
> rnorm(5,mean=17.5,sd=sqrt(3))
[1] 18.07431 18.66046 18.34520 16.51477 19.13083
```

Notice that the first three illustrations suggest ways to compute significances (p-values) for tests. The fourth illustration suggests a way of obtaining critical values for confidence intervals and tests.

The density function is rarely used except when the distribution of a data sample is to be compared with that of a theoretical distribution by graphical methods. This is illustrated later.

2.4.2   t-distributions

Let X be a continuous random variable having a t-distribution with ν degrees of freedom. Denote the density function of X by $d(x)$. Then, with notation analogous to the previous illustrations,

Function/Syntax	Computes
dt(x,ν)	$d(x)$
pt(x,ν,lower.tail=T)	$P(X \leq x)$
qt(p,ν,lower.tail=T)	x for which $P(X \leq x) = p$
rt(n,ν)	Random sample of size n

Suppose X has a t-distribution with degrees of freedom $\nu = 23$. Then, to find $P(X \leq 1.7)$, use

---

[1]Notation used for right-tailed standard normal quantiles will be $z(\alpha)$, where α represents a right-tailed probability.

```
> pt(1.7,23)
[1] 0.9486917
```

To find $P(X < -1.25$ or $X > 1.25)$, use

```
> pt(-1.25,23)+pt(1.25,23,lower.tail=F)
[1] 0.2238692
```

To find $P(-0.5 < X < 1.3)$, use

```
> pt(1.3,23)-pt(-0.5,23)
[1] 0.5858479
```

To find x for which $P(X \leq x) = 0.01$, use

```
> qt(0.01,23)
[1] -2.499867
```

and to find x for which $P(X > x) = 0.01$, use[2]

```
> qt(0.01,23,lower.tail=F)
[1] 2.499867
```

To obtain a random sample of size 5, use

```
> rt(5,23)
[1] 0.2591063 -0.2095526 0.1996334 -0.1366749 -0.4496765
```

As with normal density functions, the function dt might find use when the distribution of a data sample is to be compared with that of a theoretical distribution by graphical methods.

2.4.3   F-distributions

Let the continuous random variable X have an F-distribution with degrees of freedom ν_1 and ν_2. Let earlier notations have the same meaning with respect to probabilities and quantiles. Then

Function/Syntax	Computes
df(x,ν_1,ν_2)	$d(x)$
pf(x,ν_1,ν_2,lower.tail=T)	$P(X \leq x)$
qf(p,ν_1,ν_2,lower.tail=T)	x for which $P(X \leq x) = p$
rf(n,ν_1,ν_2)	Random sample of size n

---

[2] Notation used for right-tailed t-distributed quantiles will be $t(\alpha, df)$, where α represents a right-tailed probability and df is degrees of freedom.

Suppose X has an F-distribution with degrees of freedom $\nu_1 = 3$ and $\nu_2 = 7$. Then, to find $P(X \leq 0.5)$, use

```
> pf(0.5,3,7)
[1] 0.3059636
```

To find $P(X > 0.05)$, use

```
> pf(0.5,3,7,lower.tail=F)
[1] 0.6940364
```

To find x for which $P(X \leq x) = 0.01$, use

```
> qf(0.01,3,7)
[1] 0.03613801
```

and to find an x for which $P(X > x) = 0.01$, use[3]

```
> qf(0.01,3,7,lower.tail=F)
[1] 8.451285
```

To obtain a random sample of size 5, use

```
> rf(5,3,7)
[1] 1.1842934 1.4827499 0.8069989 2.1344250 0.3472704
```

Applications of the density function might be similar to those of earlier distributions.

2.5   User Defined Functions

The general syntax for preparing a customized function in R is as follows

```
function.name <-function(argument.list){
                R command syntax
                R command syntax
                ⋮
                R command syntax
                return(computed.value)}
```

---

[3]Notation used for right-tailed F-distributed quantiles will be $F(\alpha, df_N, df_D)$, where α represents a right-tailed probability and df_N and df_D represent degrees of freedom of the numerator and denominator, respectively.

In the previous definition, the *argument.list* can have arguments of any type (string, number, or logical). These are what the function is intended to use in its execution to `return` an answer which, in turn, may be of any type (string, number, or logical). The *argument.list* in a function definition may also be empty, written `function()`, if no external elements are needed in the function's execution.

The first brace, "{", opens the body of the function and the lines of *R command syntax* are instructions on what R is to do. The last brace, "}", indicates the end of the function. The command `return(`*computed.value*`)` is used to return the *computed.value* that the function in question is designed to compute. If this command is not present in the function body, R returns the last value that is computed. Here is a quick illustration.

Suppose X has a normal distribution, and suppose a particular set of code requires a large number of probability computations of the form $P(a \leq X \leq b)$. Given values for a and b, the mean (`mu`), and standard deviation (`s`), this would mean a large number of lines of code having the appearance

```
pnorm(b,mean=mu,sd=s)-pnorm(a,mean=mu,sd=s)
```

While this does not seem much, after retyping (or even copy-and-pasting) things may get old. In the script editor, type in exactly what is shown below and run the whole command (all lines).

```
pnorm.atob<-function(a,b,mu,s){
    return(pnorm(b,mean=mu,sd=s)-pnorm(a,mean=mu,sd=s))}
```

The console should show something like

```
> pnorm.atob<-function(a,b,mu,s){
+       return(pnorm(b,mean=mu,sd=s)-pnorm(a,mean=mu,sd=s))}
```

Remember, the "+" symbol indicates the first command line is continued onto the second. The appearance of the ">" symbol after the executed code indicates a successful compilation of the function `pnorm.atob`. On checking, the workspace will contain

```
> ls()
[1] "pnorm.atob"
```

The function `pnorm.atob` is now ready to be used.

Let X be normally distributed with mean 12.2 and standard deviation 21.7, and consider finding the probability that a randomly selected X from the population has a value that lies between 10 and 20. That is, compute $P(10 < X < 20)$. Then

```
> pnorm.atob(10,20,12.2,21.7)
[1] 0.1807462
```

This function can be saved using the dump function, and recalled whenever needed using the source function.

On a side note, if a particular package is to be used within a function, use require(*package.name*) rather than library(*package.name*). The require function is specifically designed for use inside other functions, including user-defined functions.

Chapter 3

Working with Data Structures

3.1	Introduction	35
3.2	Naming and Initializing Data Structures	36
3.3	Classifications of Data within Data Structures	38
3.4	Basics with Univariate Data	40
	3.4.1 Accessing the contents of univariate datasets	40
	3.4.2 Sorting univariate datasets	41
	3.4.3 Computations with univariate datasets	42
	3.4.4 Special vector operations	44
	3.4.5 Importing univariate datasets from external files	47
3.5	Basics with Multivariate Data	49
	3.5.1 Accessing the contents of multivariate datasets	49
	3.5.2 Computations with multivariate datasets	51
3.6	Descriptive Statistics	52
	3.6.1 Univariate datasets	53
	3.6.2 Multivariate datasets	54
3.7	For the Curious	55
	3.7.1 Constructing a grouped frequency distribution	56
	3.7.2 Using a sink file to save output	57

3.1 Introduction

In Chapter 2, computations involved *scalars* (individual numbers). Statistical data analysis generally involves what might be referred to as *vectors* (columns or lists of data), the contents of which may or may not be numerical.

This chapter looks at working with data structures that might be similar to those that are encountered in "real life" data analyses tasks. It will be found that numerical computational commands from scalars carry over nicely.

In what follows, the generic term *string (of characters)* will be used to indicate a *categorical data value*, the term *number* will be used to indicate a (real valued) *numerical data value*, and the term *logical* will be used to indicate *a Boolean (true/false) data value*. Thus, the term *data value* may refer to values from any one of these three classes and a *data structure* may refer to a collection of lists (or vectors) of data values belonging to one or more of these three classes.

3.2   Naming and Initializing Data Structures

A simplistic way of describing the name of a data structure in R is that it acts as a place-holder for the R object in question in the R workspace. Thus, the *name of a data structure* serves a purpose similar to that of a letter (for variables, constants, etc.) as used, for example, in algebra. It is important to remember that in R a name cannot be referenced unless it has been initialized or defined in some fashion.

Consider, for example, entering `ice` for the first time in a session, by itself. This action is equivalent to asking R to digest the command `ice`; then R goes ahead and looks around. The result will be:

```
> ice
Error: object 'ice' not found
```

the reason being that the object `ice` does not exist in the workspace. One way in which a *named object* may be *activated* is to use the *assignment operator* "<-" to place the data in a named object. Thus, suppose `ice` represents the proportion of a certain mountain lake, by surface area, that is covered with ice. Suppose measurements were made at the end of each month for a particular year. Then

```
> ice<-c(1,1,.8,.55,.25,.05,0,0,0,.1,.75,.95)
> ice
 [1] 1.00 1.00 0.80 0.55 0.25 0.05 0.00 0.00 0.00 0.10
[11] 0.75 0.95
```

shows that the object `ice` has been created and then recognized by R. To complete the setting for the "story" in question, create another variable and check the contents as follows.

To illustrate a command line feature, type the first line in the console exactly as it appears below and then press the Enter key. The "+" symbol at the start of the second line is R's way of saying the command from the first line is continued on the second line.

```
> month<-c("Jan","Feb","Mar","Apr","May","Jun","Jul","Aug",
+ "Sep","Oct","Nov","Dec")
> month
 [1] "Jan" "Feb" "Mar" "Apr" "May" "Jun" "Jul" "Aug"
 [9] "Sep" "Oct" "Nov" "Dec"
```

Then, the objects contained in the workspace are

```
> ls()
[1] "ice" "month"
```

Now suppose it is desired that the objects `ice` and `month` be combined in a single data structure, maybe something like a table. One might be tempted to try

```
> lake.data<-c(ice,month)
> lake.data
 [1] "1"   "1"   "0.8" "0.55" "0.25" "0.05" "0"   "0"
 [9] "0"   "0.1" "0.75" "0.95" "Jan" "Feb" "Mar" "Apr"
[17] "May" "Jun" "Jul" "Aug" "Sep" "Oct" "Nov" "Dec"
```

Clearly this does not look quite right. Two things have happened. First, all the data have been placed in a single vector (list of data values); and second, the numeric data appears to have been coerced into categorical data. On a side note, observe the use of a period in place of a space in the object name `lake.data`. Periods can be used to help add clarity to object names.

There are two functions that can be used to "bind" corresponding entries of two or more vectors of equal length. Try

```
> cbind(ice,month)
```

and

```
> rbind(ice,month)
```

It should be apparent that the "bindings" created above contain the variables `ice` and `month`, but in different formats. There is a problem, however. The data in `ice` has been coerced into character mode.

Now try

```
> lake.data<-data.frame("ice"=ice,"month"=month)
> lake.data
```

Since `ice` and `month` are already named, the above can also be accomplished using

```
lake.data<-data.frame(ice,month)
```

The output for `lake.data` now looks very similar to that of the object obtained using `cbind`, but nicer — also, `ice` retains its numerical format. *Data frames* will find a lot of use for topics covered in later chapters. We now have an example of an object that contains sub-objects and are able to illustrate one of the uses of the `names` function.

```
> names(lake.data)
[1] "ice" "month"
```

Suppose we wish to change the names of the sub-objects; then, for example,

```
> names(lake.data)<-c("Ice.Cover","Month.of.Year")
> names(lake.data)
[1] "Ice.Cover" "Month.of.Year"
```

does the trick. It is important to pay attention to the order in which the new names are assigned. Also, use periods in place of spaces in variable names, particularly if the data frame is to be used for computations at a later stage. Another caution: avoid using a numeric name for a variable, such as "2011". This appears to create problems with some R functions.

It is also possible to *initialize* a data object for later use. What this means is that a named location in the workspace is reserved for the initialized object and some default initial values are stored in the object.

For example, each of the following functions produces vectors of a given length n, each with a default value appropriate for the storage mode of the intended data.

Function/Syntax	Default Value(s)
numeric(n)	0
integer(n)	0
character(n)	""
logical(n)	FALSE
vector("*mode.type*",n)	Any of the above, depends on mode

For vector, the argument *mode.type* may be any of those recognized by R. The types that are used later include only the types logical, integer, numeric, and character.

Once a named object has been initialized (defined) by any of the approaches introduced above, the contents and names of the contents of the object can be assigned, accessed, or reassigned whenever and with whatever desired as long as any constraints imposed on the defined structure are satisfied.

3.3   Classifications of Data within Data Structures

The earlier illustrations of using c, cbind, and rbind on the ice and month data objects suggest a need for caution when combining data objects. It is important to remember that R determines the type or storage *mode* of a data list (such as a vector) according to a particular hierarchy. Of the data types (modes) that will be used later, the hierarchy is

```
character > numeric > logical
```

Thus, a list containing a string and a collection of numbers would be treated as having mode type `character`. For this reason, it is best to make sure the type, or storage mode, of data placed within a particular object using any of the functions c, cbind, and rbind is consistent. Execute each of the following commands for illustrations:

```
mode(ice)
mode(month)
mode(c(ice,month))
mode(cbind(ice,month))
mode(lake.data)
```

The function `typeof` performs a task similar to the function `mode`, the difference being that `typeof` distinguishes between integer numeric data and (non-integer) *double precision* numeric data.

In constructing data structures, it is helpful first to have a visual idea of what the structures might look like in table form. Next to scalars, the simplest type of structure is a *univariate list* of n data values. Data can also have the appearance of a (two-dimensional) table, or a collection of tables. As examples of higher dimensional data that may be encountered, consider the following scenarios:

Suppose a dataset comprises a list of three values for a response (dependent) variable Y along with corresponding values for two explanatory (independent) variables X_1 and X_2. Such data may be tabulated and stored in R in the format

Y	X_1	X_2
y_1	x_{11}	x_{12}
y_2	x_{21}	x_{22}
y_3	x_{31}	x_{32}

This is easily done using the earlier function `data.frame`. The indices for the x's are identifiers that signal to which "y-value" *and* "x-variable" a particular "x-value" belongs. For example, x_{32} is the third value of the second "x-variable", that is, the third value of X_2. So, for the x's, the first index identifies the row number and the second index identifies the column number of an "x-value".

Pushing the envelope a bit more, it may be that three such tables of data were obtained under three different treatments (or scenarios), A_1, A_2, and A_3. Then, the data might appear as follows:

A_1			A_2			A_3		
Y	X_1	X_2	Y	X_1	X_2	Y	X_1	X_2
y_{11}	x_{111}	x_{121}	y_{12}	x_{112}	x_{122}	y_{13}	x_{113}	x_{123}
y_{21}	x_{211}	x_{221}	y_{22}	x_{212}	x_{222}	y_{23}	x_{213}	x_{223}
y_{31}	x_{311}	x_{321}	y_{32}	x_{312}	x_{322}	y_{33}	x_{313}	x_{323}

There are a couple of ways in which to look at this scenario. As illustrated above, the three table objects can be given names A_1, A_2, and A_3 and matters can be taken from there, or a new categorical variable, say A, can be used in a single expanded table.

Y	y_{11}	y_{21}	y_{31}	y_{12}	y_{22}	y_{32}	y_{13}	y_{23}	y_{33}
X_1	x_{111}	x_{211}	x_{311}	x_{112}	x_{212}	x_{312}	x_{113}	x_{213}	x_{313}
X_2	x_{121}	x_{221}	x_{321}	x_{122}	x_{222}	x_{322}	x_{123}	x_{223}	x_{323}
A	A_1	A_1	A_1	A_2	A_2	A_2	A_3	A_3	A_3

This too can be stored in a data frame. In summary, the target for datasets such as those illustrated above and others is to code appropriate (additional) variables in a manner that results in a collection of lists of data (for example, Y, X_1, X_2, A) all having the same length. Then, no matter what the original tabulation format is, the coded data is always viewed as a multivariate list having the appearance of a familiar table whose columns represent variables and whose rows represent measured data values.

3.4 Basics with Univariate Data

Before beginning, clear all R objects in the current workspace by executing the command `rm(list=ls(all=TRUE))`, then use the combine function to create two univariate datasets

```
> Evens<-c(2,4,6,8,10,12)
> Colors<-c("Red","Blue","Green")
```

The first set of illustrations looks at ways to access the contents of a list of data.

3.4.1 Accessing the contents of univariate datasets

Entries of a univariate dataset are in fact a sequence of data values, much the same as a sequence of numbers and each entry can be accessed in R through the use of the entry's index.

Consider the dataset `Evens` and, in the following illustrations, execute the commands in the Syntax column to get the output in the Output column to perform the task described in the Description column.

Syntax	Output	Description
`length(Evens)`	6	Number of entries
`Evens`	2 4 6 8 10 12	List all entries
`Evens[5]`	10	Fifth entry
`Evens[-3]`	2 4 8 10 12	All but third entry
`Evens[4:6]`	8 10 12	Last three entries
`Evens[-(4:6)]`	2 4 6	All but last three entries
`Evens[seq(1,5,2)]`	2 6 10	Odd indexed entries
`Evens[Evens>4]`	6 8 10 12	Entries greater than 4

Datasets can be expanded, or entries can be changed, as long as the storage mode remains consistent. The second and third commands shown below produce no output.

Syntax	Output
`Colors`	"Red" "Blue" "Green"
`Colors<-c(Colors,"Pink")`	
`Colors[2]<-"Brown"`	
`Colors`	"Red" "Brown" "Green" "Pink"

Notice that what goes inside the *index operator* "[ ]" determines what is accessed, or changed. See `help("[")` for further details on extracting or replacing parts of an object. Also given in this documentation page are distinctions between the operators "[ ]", "[[ ]]", and "$" (to be used later).

3.4.2 Sorting univariate datasets

There are three commands that might prove useful in manipulating the order in which information in a numeric list is accessed or presented: `sort`, which sorts entries in a list in ascending or descending order; `rank`, which provides information on the alpha-numerical ranks of entries in a list in ascending order; and `order`, which provides a means of accessing a sorted list without actually rearranging the original list. Consider the following short list, and the original order of the entries.

```
> x<-c(24,23,11,24,19,23,12); x
[1] 24 23 11 24 19 23 12
```

Sorting this list in ascending order produces

```
> sort(x,decreasing=F)
[1] 11 12 19 23 23 24 24
```

The default value for `decreasing` is F, so this need not be included in the command syntax if an ascending sort is desired. If this argument is passed in as `decreasing=T`, the list is sorted in descending order.

To get the alpha-numerical ranks for the entries of x in ascending order, execute

```
> rank(x,ties.method="first")
[1] 6 4 1 7 3 5 2
```

So, the first entry in the original list is the sixth largest number, the second is the fourth largest, and so on. Other choices for the `ties.method` are `average`, `random`, `max` and `min` (which need to be enclosed within quotes when assigned); these determine how tied entries are ranked.

The function `order` provides a way to obtain the entry indices arranged in order of the sorted data without rearranging the original list. As with the function `sort`, the arrangement can be in `decreasing=F` (i.e., increasing) or in `decreasing=T` order of the entries.

```
> order(x)
[1] 3 7 5 2 6 1 4
```

To elaborate, the third entry in the unsorted list is the smallest; the seventh entry is the second smallest; the fifth entry is the third smallest; and so on. As one might expect, if we were to output the entries of x in ascending order we would get the sorted list:

```
> x[order(x)]
[1] 11 12 19 23 23 24 24
```

So, while `sort` does a physical sorting of the entries and `rank` looks at relative sizes of entries in the same order as they appear in the original list, `order` provides the (original list's) indices of the sorted version of the list in question.

Clear the R workspace before continuing.

3.4.3   Computations with univariate datasets

Operations on numerical lists work in much the same way as with numbers, but on an entry-by-entry basis with restrictions being limited to the mathematical (and sometimes R-syntax related) legality of the proposed computation.

Create two lists of the same size.

```
u<-c(1,3,-1)
v<-c(-1,0,1)
```

Then, the following arithmetic operations can be performed on these lists.

Syntax	Output	Description
u+v	0 3 0	Sum of entries
u−v	2 3 −2	Subtracts entries of v from u
2*u−3*v	5 6 −5	Computes the linear combination

Thus, the sum u+v is defined as $\{u_1 + v_1, u_2 + v_2, \ldots\}$, the difference u−v is defined as $\{u_1 - v_1, u_2 - v_2, \ldots\}$, and the product of a scalar and a vector, a*u, is defined as $\{a * u_1, a * u_2, \ldots\}$.

Similarly, in R, the product u*v is defined as $\{u_1 * v_1, u_2 * v_2, \ldots\}$ and the quotient u/v is defined as $\{u_1/v_1, u_2/v_2, \ldots\}$.

Syntax	Output	Description
u*v	−1 0 −1	Product of entries
u/v	−1 Inf −1	Division of entries
v/u	−1 0 −1	Division of entries

Note that u[2]/v[2]=Inf indicates that this entry of the resulting computation is undefined.

In general, if f represents a mathematical function, then f(u) is computed as $\{f(u_1), f(u_2), \ldots\}$.

Function/Syntax	Output	Description
round(exp(u),2)	2.72 20.09 0.37	Exponent of entries
log(v)	NaN −Inf 0	Natural log of entries
u^v	1 1 −1	Powers of entries of u

Note the warning message for log(v) (not shown above). The last operation, u^v, computes $\{u_1{}^{\wedge}v_1, u_2{}^{\wedge}v_2, \ldots\}$.

Sums and products of entries can also be obtained

Function/Syntax	Output	Description
sum(u)	3	Sum of entries of u
sum(v^2)	2	Sum of squares of entries of v
prod(u)	−3	Product of entries of u

Thus, sum computes $u_1 + u_2 + \cdots + u_n$, and prod computes $u_1 * u_2 * \cdots * u_n$.

Observe that the above illustrations involved computations with scalars (single numbers) and vectors (lists of numbers) having the same lengths. There are a couple of suggestions with respect to cases where vectors of different lengths are used in a computation. The first suggestion is *simply don't do it*. Observe what happens below. Lengthen u, and then try to add v to u.

```
> u<-c(u,2)

> u+v

[1] 0 3 0 1

Warning message:

In u + v : longer object length is not a multiple of

shorter object length
```

Here is how the (entry-wise) sums are calculated

$$1+(-1)=0 \quad 3+0=3 \quad (-1)+1=0 \quad 2+(-1)=1$$

Since v has only three entries to the four of u, R goes back to the first entry of v and adds it to the fourth entry of u.

The second suggestion is that *if you absolutely must* do such computations, make sure the length of the longer vector is a multiple of the length of the shorter vector. Lengthen u to six entries, then try adding.

```
> u<-c(3,u,1)

> u+v

[1] 2 1 4 -2 2 2
```

R has no objections. This feature is, in fact, what permits R to calculate products of the form 2*u, sums of squares such as sum(v^2), and so on. So, it is a useful feature.

In summary, if a computation makes sense in mathematical notation, it is possible to coax R into performing the operation correctly — as long as you know what you want and are careful with what you instruct R to do.

3.4.4 Special vector operations

The following discussion can be ignored without missing much in the way of overall functionality. This material applies mainly to computations that call specifically for vector and matrix operations.

For the most part, a list need not be defined specifically as a vector in order to be able to perform basic vector arithmetic. But, just in case, here are some functions that either manipulate vectors or perform operations on vectors. The vector function mentioned earlier can be used to define a vector, or a known list can be coerced into vector format as follows:

```
x<-as.vector(c(1,3,2,4))

y<-as.vector(c(-3,2,-1,3))
```

The storage mode is determined by the nature of the entries; here, the storage mode is numeric.

It appears that R stores a vector as a column vector (in the form of an $n \times 1$ matrix) rather than as a row vector (in the form of a $1 \times n$ matrix); however, caution must be exercised when performing vector operations as R seems to have the capability to think for the user and makes things work out somehow, even when a given computation is (notationally) impermissible.

To see that this really is the case, start with

```
> x
[1] 1 3 2 4
```

Next, try

```
> as.matrix(x)
     [,1]
[1,]    1
[2,]    3
[3,]    2
[4,]    4
```

The command as.matrix coerces x into a matrix and, as the output suggests, x is a column vector. The nice thing to know is that the entries of a vector can be accessed and manipulated exactly as was done for general univariate lists, and all computational tasks performed earlier on univariate lists work on vectors, too.

At times there may be a need to represent a vector as a row vector, rather than as a column vector. In such situations the transpose function, t, comes in use.

```
> t(x)
     [,1] [,2] [,3] [,4]
[1,]    1    3    2    4
```

The transpose of a vector $\mathbf{x}$ or a matrix $\mathbf{X}$ is denoted by $\mathbf{x}'$ and $\mathbf{X}'$, respectively. So, when a vector is transposed in R, the result takes on the form of a $1 \times n$ matrix. This really confirms the fact that R stores vectors as column vectors.

A common product encountered when dealing with vectors is the *dot (or inner) product*. Consider two vectors of equal length, say $\mathbf{x} = (x_1, x_2, \ldots, x_n)$ and $\mathbf{y} = (y_1, y_2, \ldots, y_n)$. The dot product of these two vectors produces a scalar, and is defined by

$$\mathbf{x} \cdot \mathbf{y} = x_1 \, y_1 + x_2 \, y_2 + \cdots + x_n \, y_n.$$

which, in R, is simply `sum(x*y)`. Observe that this product can also be obtained via matrix multiplication, if **x** and **y** are viewed as $n \times 1$ matrices, yielding a 1×1 matrix (i.e., a scalar)

$$\mathbf{x'} \, \mathbf{y} = x_1 \, y_1 + x_2 \, y_2 + \cdots + x_n \, y_n.$$

Note that the matrix product $\mathbf{x} \, \mathbf{y'}$ produces an $n \times n$ matrix,

$$\mathbf{x} \mathbf{y'} = \begin{bmatrix} x_1 \, y_1 & x_1 \, y_2 & \cdots & x_1 \, y_n \\ x_2 \, y_1 & x_2 \, y_2 & \cdots & x_2 \, y_n \\ \vdots & \vdots & \ddots & \vdots \\ x_n \, y_1 & x_n \, y_2 & \cdots & x_n \, y_n \end{bmatrix}$$

R functions available to perform these two products are `crossprod` and `tcrossprod`, respectively. Using the vectors **x** and **y**

```
> crossprod(x,y)
      [,1]
[1,]   13
> tcrossprod(x,y)
        [,1]  [,2]  [,3]  [,4]
[1,]     -3     2    -1     3
[2,]     -9     6    -3     9
[3,]     -6     4    -2     6
[4,]    -12     8    -4    12
```

Note that `tcrossprod(y,x)` will produce a matrix that is the transpose of `tcrossprod(x,y)`, and `crossprod(y,x)` produces the same result as `crossprod(x,y)`.

As long as the dimensions of both vectors are the same, the output of the associated product for both of the above functions is determined by the function used and, in the case of `tcrossprod`, the order in which the vectors are entered.

Try obtaining the product of two vectors having different lengths. What do you observe? You might have to work with simple examples, and by hand, to see what is going on.

Two other special product operators available in R are the *matrix multiplication operator* "`%*%`" for matrices and the *outer product operator* "`%o%`" for arrays of numbers. Using these functions can be risky, but they can be used to duplicate the above products *if proper care is taken*. For example, `x%*%y` duplicates `crossprod(x,y)` and `x%o%y` duplicates `tcrossprod(x,y)`.

In summary, with respect to any two vectors **x** and **y**, each having dimension n, the operation `x%*%y` produces the *inner product (or dot product)* of **x** and $\mathbf{y}$. Observe that $\mathbf{x}'$ has dimensions $1 \times n$ and $\mathbf{y}$ has dimensions $n \times 1$, so $\mathbf{x}'\mathbf{y}$ has dimensions 1×1. Similarly, the operation x%o%y produces the *outer product* of $\mathbf{x}$ and $\mathbf{y}$. Here $\mathbf{x}$ has dimensions $n \times 1$ and $\mathbf{y}'$ has dimensions $1 \times n$, so $\mathbf{x}\,\mathbf{y}'$ has dimensions $n \times n$.

Transposing either x or y in x%o%y, for example x%o%t(y), can produce unexpected results. Try this. What might be going on here?

3.4.5 Importing univariate datasets from external files

There are a few different file formats that R can import data from, two of which are demonstrated here. Clear the workspace with the function rm before continuing.

For purposes of illustration, put the data list

$$x = \{7, 3, 2, 1, 4, 2, 3, 1, 6, 5, 3, 7, 1, 8\}$$

in the first column of an Excel file (with a single sheet; delete all additional sheets), the first entry being x and the remaining being the numbers from the list. Save this file in each of the formats Data03x01.csv and Data03x01.txt, in a folder of choice, say

```
z:\Docs\RCompanion\Chapter3\Data\
```

Then, to import the dataset from the *csv* file into the unimaginatively named object SmallSet, use[1]

```
SmallSet<-read.csv(
    "z:/Docs/RCompanion/Chapter3/Data/Data03x01.csv",header=T)
```

Equivalently, to import the dataset from the *txt* file, use

```
SmallSet<-read.table(
    "z:/Docs/RCompanion/Chapter3/Data/Data03x01.txt",header=T)
```

In either case, the data are stored in the object SmallSet, which is a data frame and has the storage mode of list. On typing SmallSet, a table of two columns will appear on the screen, the first column being the indices or data entry position numbers and the second being the data column having name x. If the objects contained in the workspace are listed, execute ls(); it will be noticed that the only object present is SmallSet. Also,

```
> names(SmallSet)
[1] "x"
```

---

[1] Enter the following on a single line in the script editor.

suggests that the variable x is *inside* the data frame SmallSet. There are four ways in which the contents of SmallSet (the numbers in x for this case) can be accessed. All of the following command lines produce the same output:

```
SmallSet[,1]

SmallSet$x

attach(SmallSet); x; detach(SmallSet)

with(SmallSet,x)
```

The first approach instructs R to extract all rows in the first object (x) in the list of objects (SmallSet). The second approach is equivalent, but is more intuitive in that R is instructed to extract the object x contained in SmallSet, the symbol "$" indicates containment. The third approach attaches (the contents of) SmallSet to the workspace for direct access. After being used, the attached object is detached so as to restrict access to its contents. The last approach can be thought of as a safe alternative to attaching SmallSet; the tendency is usually to forget to detach an object after it has been attached. The with function ensures that the contents of SmallSet remain restricted after use.

Once any data object, such as SmallSet, has been created in the workspace, it can be saved using the dump function. The source code for the data object can then be reloaded at a later date using the source function. As an illustration,

```
> dump("SmallSet",

+ "Z:/Docs/RCompanion/Chapter3/Data/Data03x01.R")

> rm(list=ls(all=T));ls()

character(0)

> source("Z:/Docs/RCompanion/Chapter3/Data/Data03x01.R")

> ls()

[1] "SmallSet"
```

See help(data,package=utils) and help(save,package=base) for additional options available in this area.

For convenience, the contents of x can be brought out and stored in an external variable using a command of the form

```
x<-SmallSet$x; x
```

Clear all objects in the workspace before continuing.

3.5 Basics with Multivariate Data

Consider the data shown in Table 3.1 where A_1, A_2, and A_3 represent levels of a factor (catagorical variable), Y represents a continuous response variable, and X represents a continuous explanatory variable (covariate).

Use Excel to prepare this data for importing it into R. In the first row enter names Y, X, and A. Under Y place all y-values moving in columns from left to right. In the second column under X place all the x-values, again moving from left to right. Finally, under A, fill in the category number to which each xy-pair belongs; place 1 in the first four cells, place 2 in the next four cells, and place 3 in the last four cells. The result will be a table with three columns, a header row, and twelve rows of data.

TABLE 3.1: Data for Section 3.5 Illustrations

A_1		A_2		A_3	
Y	X	Y	X	Y	X
32	0	5	21	3	12
43	15	32	75	35	48
72	21	42	87	70	55
95	42	82	95	90	82

Save this file as a "`*.txt`" (or "`*.csv`") file, then read it using the `read.table` (or `read.cvs`) command. For example,

```
CatSet<-read.table(
    "z:/Docs/RCompanion/Chapter3/Data/Data03x02.txt",header=T)
```

stores the contents of `Data03x02.txt` in the data frame `CatSet`. As demonstrated for `SmallSet`, the data frame `CatSet` can be saved using the `dump` function,

```
dump("CatSet",
    "Z:/Docs/RCompanion/Chapter3/Data/Data03x02.R")
```

and reloaded using the `source` function.

3.5.1 Accessing the contents of multivariate datasets

Access to the contents of `CatSet` can be achieved in much the same manner as for univariate data frames, with the added need to specify which object in the data frame is of interest. Simply executing `CatSet` will list the contents of `CatSet` in table form (not shown). The contents may be explored as before:

```
> names(CatSet)
[1] "Y" "X" "A"
> CatSet$Y; CatSet$X; CatSet$A
[1] 32 43 72 95 5 32 42 82 3 35 70 90
[1] 0 15 21 42 21 75 87 95 12 48 55 82
[1] 1 1 1 1 2 2 2 2 3 3 3 3
```

As with the univariate dataset from earlier, the `attach` function can be used to make the contents of `CatSet` directly accessible, and the `with` function serves the same purpose more elegantly and safely.

As with univariate datasets, Boolean statements can be used to look for specific characteristics of a variable. For example, to list all values above 35 within the variable Y, execute

```
> CatSet$Y[CatSet$Y>35]
[1] 43 72 95 42 82 70 90
```

or, more simply, using

```
> with(CatSet,Y[Y>35])
[1] 43 72 95 42 82 70 90
```

Other commands applicable to univariate datasets apply to each of the variables within a multivariate data structure such as `CatSet`.

Factors, or *categorical variables* within a data structure should be defined prior to performing any quantitative procedures. For `CatSet` the variable `A` has numeric entries, but is not meant to be treated as a numerical variable. To define `A` as being categorical, execute

```
CatSet$A<-factor(CatSet$A)
```

Levels, or distinct categorical values within a factor may be listed and reassigned if desired. For example,

```
> levels(CatSet$A)
[1] "1" "2" "3"
> levels(CatSet$A)<-c("A1","A2","A3")
> with(CatSet,levels(A))
[1] "A1" "A2" "A3"
```

Note the two equivalent approaches used to list the levels of `A`. As with univariate data, it will be necessary to be able to work on the contents of multivariate data structures.

Another approach to looking at the contents of CatSet might be as follows: Suppose we wished to see all entries for which Y>35, rather than just the entries of Y. Then,

```
> with(CatSet,CatSet[Y>35,])
   Y  X   A
2  43 15 A1
3  72 21 A1
4  95 42 A1
7  42 87 A2
8  82 95 A2
11 70 55 A3
12 90 82 A3
```

The comma in `CatSet[Y>35,]` without anything following tells R to look at all columns. Similarly,

```
> with(CatSet,CatSet[(X>0)&(X<55),c(2,3)])
   X  A
2  15 A1
3  21 A1
4  42 A1
5  21 A2
9  12 A3
10 48 A3
```

instructs R to list all rows in columns 2 and 3 for which X>0 and X<55.

One could use the `dump` function to save the altered `CatSet` for later use, for example,

```
dump("CatSet",
     "Z:/Docs/RCompanion/Chapter3/Data/Data03x02.R")
```

Remove all objects from the workspace before continuing.

3.5.2   Computations with multivariate datasets

Once a multivariate structure such as `CatSet` is created in R, computation methods for each numerical vector within the structure work as for earlier described univariate datasets. The only new set of computations that will be encountered are those involving matrix operations. Here is an illustration.

```
X<-matrix(c(1,1,1,1,3,0,-1,2),nrow=4,ncol=2,byrow=F)
```

and then enter X. It will be found that the contents of X have the following format:

```
> X
      [,1] [,2]
[1,]    1    3
[2,]    1    0
[3,]    1   -1
[4,]    1    2
```

This stores the matrix $\mathbf{X}$, shown above, in the workspace. Then, to obtain the product $\mathbf{X'X}$, execute

```
> XTX<-crossprod(X,X); XTX
      [,1] [,2]
[1,]    4    4
[2,]    4   14
```

Note that the command

```
t(X)%*%X
```

also performs this operation. To obtain the inverse of a non-singular matrix, the function `solve` is used. Thus, to obtain $(\mathbf{X'X})^{-1}$ execute

```
> XTX.Inv<-solve(XTX); XTX.Inv
       [,1]    [,2]
[1,]   0.35   -0.1
[2,]  -0.10    0.1
```

If a confirmation of the result is desired,

```
> round(crossprod(XTX,XTX.Inv))
      [,1] [,2]
[1,]    1    0
[2,]    0    1
```

shows that the product of $(\mathbf{X'X})$ and $(\mathbf{X'X})^{-1}$ is the identity matrix, approximately, as a consequence of round-off errors. As usual, clear the workspace before continuing.

3.6 Descriptive Statistics

This section summarizes some of the capabilities of R in obtaining typical descriptive statistics and data summaries. For the most part, ideas and methods from univariate datasets extend to multivariate data.

3.6.1 Univariate datasets

Consider the data sample for a continuous random variable X shown in Table 3.2. These data can be imported from a text file, say Data03x03.txt, using the earlier described read.table command. Name the resulting data frame UniVarSet. For example, the commands

```
UniVarSet<-read.table(
    "z:/Docs/RCompanion/Chapter3/Data/Data03x03.txt",header=T)
x<-UniVarSet$x; rm(UniVarSet)
```

result in the workspace having only the object x.

TABLE 3.2: Data for Section 3.6.1 Illustrations

1	5	35	40	40	40	43	43	43	45	48	50
51	51	51	51	52	52	53	54	54	55	56	56
60	61	61	61	61	61	61	67	67	67	67	67
68	68	68	68	68	69	69	70	70	70	70	70
71	75	75	75	75	76	76	76	76	76	77	77
78	78	81	81	81	81	81	81	81	85	85	85
87	87	88	88	88	88	89	89	90	90	90	90
92	93	94	94	94	94	95	95	96	101	101	102
106	107	50	60	68	71	77	86	91	105		

Here is a list of common functions that might find use in describing the data.

Function/Syntax	Output	Description
dim(x)	NULL	Dimensions of x
length(x)	106	Number of entries
mean(x)	71.48113	Sample mean
var(x)	381.833	Sample variance
sd(x)	19.54055	Sample standard deviation
min(x)	1	Minimum
max(x)	107	Maximum
range(x)	1 107	Minimum and maximum
median(x)	73	Median
quantile(x,.25)	60.25	First quartile
IQR(x)	26.75	Inter-quartile range

The commands (output not shown)

```
summary(x); fivenum(x)
```

produce summary statistics and the five-number summary.

3.6.2   Multivariate datasets

Consider the dataset contained in Table 3.3. Set the data up in a text file and import it into a data frame, say MultiNum, using the read.table function.

TABLE 3.3:   Data for Section 3.6.2 Illustrations

y	7.0	5.0	6.2	5.2	6.2	5.2	6.2
x_1	51.5	41.3	36.7	32.2	39.0	29.8	51.2
x_2	19	14	11	13	13	12	18
x_3	0.103	0.046	0.056	0.067	0.075	0.049	0.073

y	6.4	6.4	6.4	5.4	6.4	5.4	6.7
x_1	46.0	61.8	55.8	37.3	54.2	32.5	56.3
x_2	14	20	17	13	21	11	19
x_3	0.061	0.066	0.081	0.047	0.083	0.061	0.083

y	6.7	5.9	6.9	6.9	6.9	7.9
x_1	52.8	47.0	53.0	50.3	50.5	57.7
x_2	17	15	16	16	14	22
x_3	0.085	0.052	0.085	0.069	0.074	0.120

Quick information about the names of the contents of MultiNum can be obtained using

```
> names(MultiNum)
[1] "y" "x1" "x2" "x3"
```

and the dimensions of the data frame can be obtained using

```
> dim(MultiNum)
[1] 20 4
```

Summary statistics of the included variables can be obtained.

```
> summary(MultiNum)
       y                x1              x2              x3
 Min.   :5.000   Min.   :29.80   Min.   :11.00   Min.   :0.04600
 1st Qu.:5.775   1st Qu.:38.58   1st Qu.:13.00   1st Qu.:0.05975
 Median :6.400   Median :50.40   Median :15.50   Median :0.07100
 Mean   :6.265   Mean   :46.84   Mean   :15.75   Mean   :0.07180
 3rd Qu.:6.750   3rd Qu.:53.30   3rd Qu.:18.25   3rd Qu.:0.08300
 Max.   :7.90    Max.   :61.80   Max.   :22.00   Max.   :0.12000
```

The `with` function or the `$` operator can be used to single out specific variables; for example,

```
> var(MultiNum$y)
[1] 0.5466053
> with(MultiNum,fivenum(x2))
[1] 11.0 13.0 15.5 18.5 22.0
```

The earlier `var` function can be applied to a multivariate dataset, too. Since the data frame contains only numerical data, the *variance-covariance matrix* for the variables can be obtained

```
> round(var(MultiNum),3)
        y      x1      x2     x3
y   0.547   5.403   1.643  0.012
x1  5.403  89.594  27.354  0.113
x2  1.643  27.354  10.829  0.046
x3  0.012   0.113   0.046  0.000
```

An alternative, but equivalent, function to obtain the variance-covariance matrix is `cov`. The *correlation coefficient matrix* can also be obtained

```
> round(cor(MultiNum),3)
        y      x1      x2     x3
y   1.000   0.772   0.675  0.838
x1  0.772   1.000   0.878  0.635
x2  0.675   0.878   1.000  0.736
x3  0.838   0.635   0.736  1.000
```

Pearson's method is the default method used for all three of these functions. See the R documentation page for details on additional arguments; documentation for all three functions is on the same page.

3.7 For the Curious

The content of this section is not needed for later chapters; however, the material given may be of interest to some.

3.7.1   Constructing a grouped frequency distribution

Recall the variable x extracted earlier from UniVarSet and consider constructing code to produce a grouped frequency distribution table of the data. Enter the given code in the script editor and run the lines in the order shown.

First find the range of the data.

```
> range(x)
[1] 1 107
```

Using a lower bound of 0 and an upper bound of 110, store boundary points for classes of width 10 in a variable and use the R function cut to partition the data into 11 classes enclosed by the boundary points.

```
bpts<-seq(0,110,10)
class<-cut(x,breaks=bpts,include.lowest=T,right=F)
```

Next, obtain the midpoints, frequencies, and relative frequencies for each class.

```
mp<-seq(5,105,10);f<-summary(class);rf<-round(f/sum(f),3)
```

Finally, store the results in a data object using the function cbind.

```
freq.dist<-cbind("Mid Point"=mp,"Freq."=f,"Rel. Freq."=rf)
```

Then, the contents of freq.dist can be seen by executing

```
> freq.dist
          Mid Point  Freq.  Rel. Freq.
[0,10)            5      2       0.019
[10,20)          15      0       0.000
[20,30)          25      0       0.000
[30,40)          35      1       0.009
[40,50)          45      8       0.075
[50,60)          55     14       0.132
[60,70)          65     21       0.198
[70,80)          75     21       0.198
[80,90)          85     19       0.179
[90,100)         95     14       0.132
[100,110]       105      6       0.057
```

The row names are inherited from the list f. The same task as cbind can be accomplished using data.frame.

The above steps can be placed in a function. However, to make the function useful and broadly applicable, quite a bit more control will have to be given to R in making decisions. Chapter 5 covers this area of programming with R.

3.7.2 Using a sink file to save output

The `sink` function might sometimes be useful in exporting large quantities of output into a text file for later analysis. As an illustration, consider placing the contents of `freq.dist` in a text file.

Start by opening an output file in the appropriate directory; for example,

```
sink(file="Z:/Docs/RCompanion/Chapter3/Data/freqdist.txt",
                                   append=F,type="output")
```

Now, every time a command is entered, the output for the command will be redirected to the file `freqdist.txt`. Then, executing

```
freq.dist
```

prints the output to the file `freqdist.txt`. If `append=T` is used, all output is appended to existing output in the output file. Having `append=F` instructs R to rewrite the output file. To close the sink file, execute

```
sink()
```

Access to the contents of `freqdist.txt` can be gained with the help of any text editor. As run above, nothing shows up on the console, so it may be inconvenient when the tracking of computational results is important. If tracking computational results is important, include the argument `split=T` in the `sink` function. This causes output to also be directed to the screen.

Chapter 4

Basic Plotting Functions

4.1   Introduction 59
4.2   The Graphics Window 61
4.3   Boxplots 61
4.4   Histograms 63
4.5   Density Histograms and Normal Curves 67
4.6   Stripcharts 69
4.7   QQ Normal Probability Plots 70
4.8   Half-Normal Plots 72
4.9   Time-Series Plots 73
4.10 Scatterplots 75
4.11 Matrix Scatterplots 76
4.12 Bells and Whistles 76
4.13 For the Curious 78
    4.13.1 QQ normal probability plots from scratch 78
    4.13.2 Half-normal plots from scratch 79
    4.13.3 Interpreting QQ normal probability plots 80

4.1   Introduction

This chapter first outlines some basic plotting functions that are used in later chapters, then covers some methods by which figures can be enhanced. For the most part, graphing functions used in this chapter belong to package `graphics` or package `grDevices` [56].

Before beginning, clear the workspace and the console and generate some data to use in the illustrations presented in this chapter. Enter the following code in the script editor and run it. Lines beginning with the "#" symbol are comments lines, added for explanation

```
#Generate a random sample in three groups
y<-c(round(rnorm(500,25,8),1),
    round(rnorm(350,15,4.5),1),round(rnorm(700,7,2.1),1))
#Create a variable to identify groups in y
G<-factor(c(rep(1,500),rep(2,350),rep(3,700)))
```

To make the data sound more interesting, suppose the data represent weights measured for random samples of three salmon species: King salmon, Coho

salmon, and Keta salmon caught in Southeast Alaska waters. Start by storing
y and G in a data frame and then remove y and G from the workspace.

```
Salmon<-data.frame(y,G); rm(y,G)
```

Next, rename the variables in `Salmon`, and then assign names to the three
levels of `Species`

```
names(Salmon)<-c("Weight","Species")
levels(Salmon$Species)<-c("King","Coho","Keta")
```

Now check to see what the data look like by executing `Salmon` to make sure
there are no negative weights. If a negative weight shows up in the summary,
re-run the code to generate a new set. A variety of data summaries can be
obtained. For an overall summary, use

```
> summary(Salmon)
     Weight        Species
 Min.   : 4.50   King:500
 1st Qu.: 7.60   Coho:350
 Median :12.10   Keta:700
 Mean   :14.87
 3rd Qu.:22.30
 Max.   :36.20
```

which summarizes the contents of the two variables in `Salmon`. For a breakdown by species, use

```
> with(Salmon,tapply(Weight,Species,summary))
$King
   Min. 1st Qu.  Median    Mean 3rd Qu.    Max.
  14.90   22.40   25.40   25.07   27.63   36.20
$Coho
   Min. 1st Qu.  Median    Mean 3rd Qu.    Max.
   4.70   12.50   15.05   14.95   17.40   24.30
$Keta
   Min. 1st Qu.  Median    Mean 3rd Qu.    Max.
  4.500   6.800   7.500   7.531   8.200  11.400
```

which summarizes `Weight` by each level in `Species`. If all looks well (no negative weights), save the dataset for later use. For example,

```
dump("Salmon",

        "Z:/Docs/RCompanion/Chapter4/Data/Data04x01.R")
```

Check to see that the data frame was indeed saved in the named folder. Recall, to reload this dataset, execute

```
source("Z:/Docs/RCompanion/Chapter4/Data/Data04x01.R")
```

4.2   The Graphics Window

The syntax for opening a graphics window with specific dimensions and a defined point size for the text in the window is[1]

```
win.graph(width=x,height=y,pointsize=n)
```

where x and y are in inches and n is a positive integer (the default `pointsize` is 12). Further adjustments to a graphics window can be made by defining various graphics parameters through the function `par`, which is contained in package `graphics`. Examples of simpler adjustments appear in the following sections; see the R documentation pages of `win.graph` and `par` for further details.

4.3   Boxplots

The function for producing boxplots is called `boxplot`. This function is quite flexible in what it takes in for data.

Consider, for example, constructing a boxplot of only `Weights` from the simulated data. The `with` function might be preferred here (over the `attach` function or the `$` operator).

```
#Set a graphics window size and point size

win.graph(width=4,height=2.5,pointsize=10)

#Obtain the boxplot of Weight with Salmon data

with(Salmon,boxplot(Weight,horizontal=T,xlab="Weight"))
```

---

[1] **Mac Alert**: Mac users, the function `quartz` is used in place of `win.graph`. See help pages on this.

The result is shown in Figure 4.1. The default value for the argument `horizontal` is `False`, so if this is left out of the `boxplot` command, a vertically oriented boxplot is displayed. The boxplot of `Weight` for an individual species, say `King` can be obtained (see Figure 4.2) using the following code.

```
#Set a graphics window size and point size
win.graph(width=4,height=2.5,pointsize=10)
#With Salmon, obtain the boxplot of weight for Kings only
with(Salmon,boxplot(Weight[Species=="King"],
        horizontal=T,xlab="Weights of Kings"))
```

Finally, boxplots for all three species can be placed in a single figure by instructing R to plot `Weight` by `Species` (see Figure 4.3).

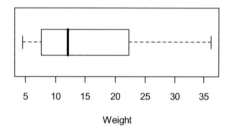

FIGURE 4.1: Boxplot of weights of all salmon species given in dataset Salmon.

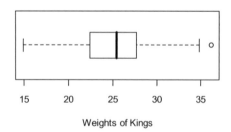

FIGURE 4.2: Boxplot of only King salmon weights extracted from the dataset Salmon.

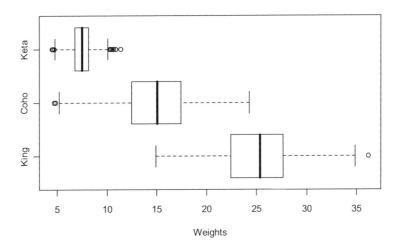

FIGURE 4.3: Boxplots of weight for each species placed in a single graphics image.

```
win.graph(width=6,height=4,pointsize=10)
#Plot weights by species
with(Salmon,boxplot(Weight~Species,
                horizontal=T,xlab="Weights"))
```

Some suggestions on enhancing figures are provided in the last section of this chapter.

4.4   Histograms

Density histogram equivalents of Figures 4.1 and 4.2 can be obtained using the function `hist` (see Figure 4.4).

```
win.graph(width=4,height=3,pointsize=10)
#Plot the density histogram of overall weights
with(Salmon,hist(Weight,freq=F))
```

The default value for the argument `freq` is `TRUE`, so if this is left out of a histogram command, a frequency histogram is constructed.

Similarly, a frequency histogram of only the weights of Kings (see Figure 4.5) can be obtained using

```
win.graph(width=4,height=3,pointsize=10)
#Plot the frequency histogram of King salmon weights
with(Salmon,hist(Weight[Species=="King"],
    xlab="Weights",main="Histogram for King Salmon"))
```

Observe that the argument `main` provides a way to customize the main title of a figure. To suppress the assignment of a default main title for a histogram, set `main=NULL`.

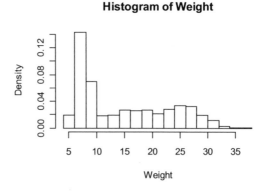

FIGURE 4.4: Density histogram of weights for all species from dataset Salmon.

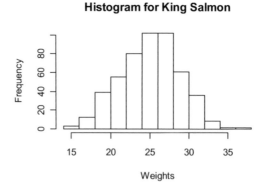

FIGURE 4.5: Frequency histogram for only King salmon weights in dataset Salmon.

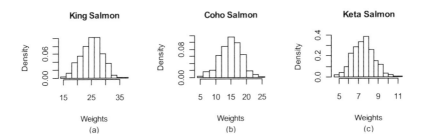

FIGURE 4.6: Separate density histograms for each species from dataset Salmon.

Unlike the function `boxplot`, the function `hist` does not have a formula capability in which histograms of weights by species can be obtained. However, a feature of the graphics device window allows the plotting of more than one figure within a single window. The trick is to partition the graphics window into sub-windows of equal area (see Figure 4.6 for the result of executing the following).

```
#Open a suitably sized window for three side-by-side plots
win.graph(width=6,height=2)
#Partition the graphics window into one row with three columns
par(mfrow=c(1,3),ps=10)
#Plot each of the density histograms
with(Salmon,hist(Weight[Species=="King"],
    xlab="Weights",freq=F,main="King Salmon",sub="(a)"))
with(Salmon,hist(Weight[Species=="Coho"],
    xlab="Weights",freq=F,main="Coho Salmon",sub="(b)"))
with(Salmon,hist(Weight[Species=="Keta"],
    xlab="Weights",freq=F,main="Keta Salmon",sub="(c)"))
#Reset the graphics window
par(mfrow=c(1,1))
```

Be aware that it is the `mfrow` argument passed into the graphics parameters function `par` that does the partitioning. The first entry in `c(1,3)` instructs R to give the graphics window one row, and the second instructs R to partition the row into three columns. The other argument, `ps`, sets the point size of the text in the figure.

On viewing the R documentation page for the function `hist`, it will be noticed that R has some options for how class widths for histograms are determined. The boundary points, or break points, between classes can be determined by R automatically through some optional algorithm, or can be user defined.

The default setting for determining break points between histogram classes is `breaks="Sturges"`. If n is the number of data values, the Sturges algorithm divides the data range into approximately $\log_2(n) + 1$ classes and attempts to place break points on "nice" numbers. Notice that this will not work too well for very large datasets as it may give too few classes. For example, with the King sample, which has 500 observations, the number of classes that might be expected is around $\log_2(500) + 1 \approx 10$. The other two options include `breaks="Scott"`, which computes the number of break points based on the standard error of the data, and `breaks="Freedman-Diaconis"`, which computes the number of break points based on the interquartile range of the data.

User-specified break points may be passed into the function `hist` by providing a single number specifying the number of bins to use, or a vector of boundary points starting with the lower bound and ending with the upper bound. Figure 4.7 shows four density histograms for King salmon weights obtained using the following code:

```
#Open a graphics window, partition into two rows and columns
win.graph(width=6,height=5); par(mfrow=c(2,2),ps=10)
#Plot histogram using Scott's algorithm
with(Salmon,hist(Weight[Species=="King"],freq=F,
    xlab="Weights",breaks="Scott",main="Scott's Algorithm"))
#Plot histogram using the Freedman-Diaconis algorithm
with(Salmon,hist(Weight[Species=="King"],freq=F,
    xlab="Weights",breaks="Freedman-Diaconis",
                    main="Freedman-Diaconis Algorithm"))
#Plot histogram using six bins
with(Salmon,hist(Weight[Species=="King"],freq=F,
        xlab="Weights",breaks=6,main="Six Bins Specified"))
#Plot histogram using given boundary points, 10,12,14,...,40
with(Salmon,hist(Weight[Species=="King"],freq=F,
                xlab="Weights",breaks=seq(10,40,2),
                main="Specified Boundary Points"))
#Reset graphics window
par(mfrow=c(1,1))
```

At times, one might wish to compare the shape of a sample's distribution with that of a theoretical distribution. A graphical approach to this is to superimpose a density curve of the desired theoretical distribution on the density histogram of the sample.

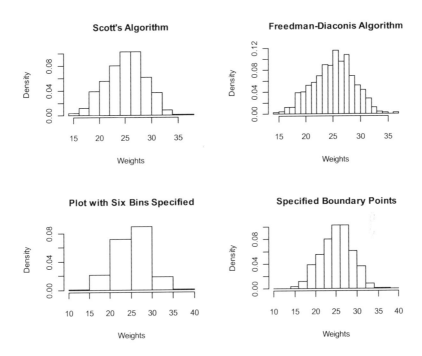

FIGURE 4.7: Density histograms of King salmon weights obtained using alternative ways to set class widths for bars.

4.5   Density Histograms and Normal Curves

Consider the sample weights for King salmon, again. Here some additional graphical features will be used to enhance the resulting figure. The first task might be to find out a bit about the data.

```
> with(Salmon,summary(Weight[Species=="King"]))

 Min.   1st Qu.  Median    Mean  3rd Qu.    Max.

14.90    22.40   25.40    25.07   27.63    36.20
```

A lower bound of 14 and an upper bound of 38 with class boundaries at $\{14, 16, 18, \ldots, 36, 38\}$ for weights might work well. Looking back at Figure 4.6.a, it appears that a safe upper limit for the density is 0.12.

Begin the task by first deciding upon a graphics window size, and then prepare the density histogram using the above horizontal and vertical axes ranges.

```
#Open a window
win.graph(width=4,height=3.5,pointsize=10)
#Specify x and y limits
x.lims<-c(14,38); y.lims<-c(0,.12)
#Specify break points
b.pts<-seq(14,38,2)
#Plot the density histogram
with(Salmon,hist(Weight[Species=="King"],freq=F,
#without axes with given x and y limits
        axes=F,xlim=x.lims,ylim=y.lims,breaks=b.pts,
#and without any titles or axes labels
        xlab=NULL,ylab=NULL,,main=NULL))
```

Next, prepare and superimpose the normal density curve onto the histogram.

```
#Obtain a list of at least 100 x-values
x.vals<-seq(14,38,.24)
#Obtain the mean and standard deviations for King weights
mu<-with(Salmon,mean(Weight[Species=="King"]))
s<-with(Salmon,sd(Weight[Species=="King"]))
#Obtain corresponding theoretical densities
y.vals<-dnorm(x.vals,mean=mu,sd=s)
#Superimpose the density curve on the histogram
lines(x.vals,y.vals,lty=2)
```

Now, touch up the figure with axes

```
#Put in the x-axis, 1, and y-axis, 2
axis(1,seq(14,38,2)); axis(2,seq(0,.12,.03))
```

and desired titles and axes labels

```
#Put in the figure title, a subtitle, and axes labels
title(main="Density histogram and density curve",
    sub="King Salmon Sample",xlab="Weight",ylab="Density")
```

The result is shown in Figure 4.8. The function `lines` plots line segments between consecutive pairs (plotted points) from `x.vals` and `y.vals`. Having a large number of such points gives the appearance of a smooth curve.

Density histogram and density curve

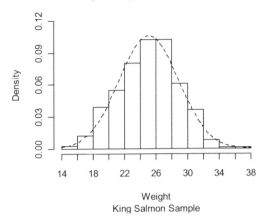

FIGURE 4.8: Density histogram of King salmon weights with superimposed normal density curve constructed using data statistics.

4.6 Stripcharts

An alternative to boxplots such as Figure 4.3, "jittered" stripcharts provide a bit more information about how individual data values are spread (see Figure 4.9). The code for this plot includes some plot formatting features mentioned earlier.

```
#Open a window and scale plotted points by factor of .75
win.graph(width=5,height=3); par(cex=.75)
#Plot a vertical stripchart of weight against species
with(Salmon,stripchart(Weight~Species,vertical=T,
#Plot circles, jittered by an amount of .25
    pch=1,method="jitter",jitter=.25))
```

The default point size for the graphics window is 12; `cex=.75` scales everything in the figure by a factor of 0.75 of the preset size. It appears this is the only way to scale the size of the plotted points, the difficulty being that it also scales the font size. The argument `pch` instructs R on which symbol to use for plotted points.

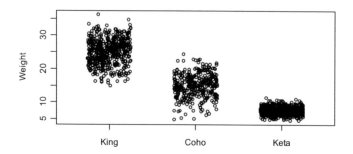

FIGURE 4.9: A jittered stripchart of weights separated by species for dataset Salmon.

4.7   QQ Normal Probability Plots

QQ normal probability plots are used to compare sample data against a theoretical normal distribution. For the three samples of salmon weight data, QQ plots can be obtained in a manner that might allow a comparison of the three (see Figure 4.10).

```
#Open a window
win.graph(width=6,height=2.5)
#Format the window and set the scaling factor
par(mfrow=c(1,3),cex=.75)
#Obtain common x- and y-limits for the axes
x.lims<-c(-4,4);y.lims<-c(0,max(Salmon$Weight))
#Plot the three QQ normal probability plots
with(Salmon,qqnorm(Weight[Species=="King"],
    xlim=x.lims,ylim=y.lims,sub="King Salmon",main=NULL))
with(Salmon,qqnorm(Weight[Species=="Coho"],
    xlim=x.lims,ylim=y.lims,sub="Coho Salmon",main=NULL))
with(Salmon,qqnorm(Weight[Species=="Keta"],
    xlim=x.lims,ylim=y.lims,sub="Keta Salmon",main=NULL))
#Reset window settings, just in case
par(mfrow=c(1,1),ps=12,cex=1)
```

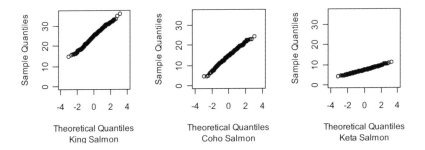

FIGURE 4.10: QQ normal probability plots for the weights for three salmon species in dataset `Salmon`.

For the remaining illustrations, extract the Coho data for easy access.

```
Coho.wts<-with(Salmon,Weight[Species=="Coho"])
```

In interpreting a QQ normal probability plot, it is helpful to include a reference line. One commonly used reference line is that which passes through the first and third quartiles of the data.

```
win.graph(width=3,height=3); par(ps=10,cex=.75)
qqnorm(Coho.wts,main=NULL,sub="Coho Weights")
#Insert the QQ line for reference
qqline(Coho.wts,lty=3)
```

Figure 4.11 contains the resulting plot. Note that the argument `lty=3` in the function call for `qqline` instructs R to draw a dashed line.

There are many who prefer to use standardized data in QQ normal probability plots (see Figure 4.12). This is easily accomplished as follows.

```
#First standardize the Coho data
S.Coho.wts<-(Coho.wts-mean(Coho.wts))/sd(Coho.wts)
#Then plot the QQ plot²
qqnorm(S.Coho.wts,main=NULL,sub="Standardized Coho Weights")
#Now plot the line y=x as the reference line
abline(a=0,b=1,lty=3)
```

The function `abline`, as shown here, plots a line having a y-intercept of a and a slope of b.

---

[2]From here on the graphics window formatting functions `win.graph` and `par` will not be shown; however, it should be understood these are used for every figure presented in all that follows.

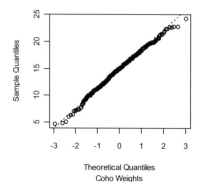

FIGURE 4.11: QQ normal probability plot of Coho weights with superimposed `qqline` for reference.

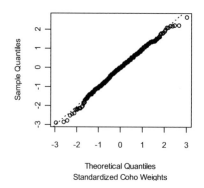

FIGURE 4.12: QQ normal probability plot of standardized Coho weights with the line $y = x$ superimposed for reference.

4.8   Half-Normal Plots

While QQ normal probability plots serve mainly to assess the normality of the underlying random variable for a univariate sample, a variation of this plot is used to highlight outlying data in a sample. A *half-normal plot* for a sample involves plotting the ordered absolute values (usually of residuals, leverages,

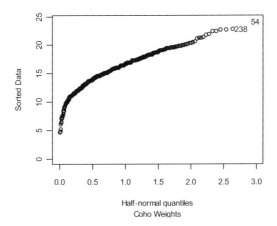

FIGURE 4.13: Half-normal plot of Coho salmon weights using the function `halfnorm` from package `faraway`.

or other model derived data used in outlier analysis) against a corresponding list of (positive) standard normal quantiles (see Figure 4.13).

Package `faraway` [23] contains a function, `halfnorm`, that plots a half-normal plot. If package `faraway` is installed on your computer, then, to load package `faraway` and execute the function `halfnorm` using `Coho.wts`, run the code

```
library(faraway)
#Plot the half-normal plot
halfnorm(Coho.wts,nlab=2)
#Put in a subtitle
title(sub="Coho Weights")
detach(package:faraway)
```

The `nlab=2` in function `halfnorm` instructs R to label the two most extreme plotted points.

4.9 Time-Series Plots

A *time-series plot* for a dataset, say $\{x_1, x_2, \ldots x_n\}$ typically observed over equal intervals of time, involves plotting the set of ordered pairs having the form $\{(i, x_i) \mid i = 1, 2, \ldots, n\}$. The horizontal axis represents the data position

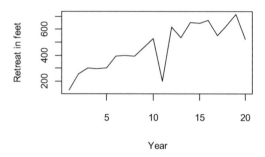

FIGURE 4.14: A basic time-series plot of the glacier retreat data from x using the function `plot.ts`.

or time order in which it has been entered, and the vertical axis represents the variable associated with the data. The R function for obtaining a time-series plot is `plot.ts`.

For example, suppose annual measurements of a certain glacier, starting from 1991 through to 2010, showed the glacier retreating by the following number of feet each year:

$$130, \ 255, \ 301, \ 295, \ 305, \ 395, \ 400, \ 395, \ 463, \ 530,$$
$$200, \ 620, \ 532, \ 650, \ 645, \ 670, \ 551, \ 632, \ 713, \ 523.$$

Then, the data can be stored in a variable, say `x`,

```
x<-c(130,255,301,295,305,395,400,395,463,530,
     200,620,532,650,645,670,551,632,713,523)
```

and then plotted

```
plot.ts(x,xlab="Year",ylab="Retreat in feet")
```

The result is shown in Figure 4.14. The function `ts` can be used to store the glacier retreat data as a time-series as follows.

```
Retreat<-ts(x,start=1991,end=2010,frequency=1)
```

Now, plot `Retreat` using each of the commands `plot(Retreat)` and `plot.ts(Retreat)` and compare the resulting figures (not shown) with Figure 4.14.

4.10   Scatterplots

The `plot` function just used is quite versatile. For example,

```
with(Salmon,plot(Weight~Species,horizontal=T))
```

produces Figure 4.3. Execute `methods(plot)` for a list of methods available to the function `plot`. In earlier illustrations, the methods `plot.ts` and `plot.factor` were used; in what follows, only the `plot.default` and `plot.formula` methods are presented where both the *independent* (*explanatory*) variable and the *dependent* (*response*) variable are considered continuous. Plots of the response (denoted by y here) against the explanatory variable (denoted by x) can be obtained using either of the commands `plot(x,y)` or `plot(y~x)`

Most arguments that can be passed into the plot function have been seen in earlier plotting functions and are listed below.

Argument	Purpose
`type="`*style*`"`	Determines the plot style, typically: lines (`"l"`), points (`"p"`), or both (`"b"`)
`main="`*main title*`"`	Assign main title for figure. Using `NULL` leaves this blank
`sub="`*Subtitle*`"`	Assigns subtitle, ignores if left out
`xlab="`*x label*`"`	Assigns x-axis label, uses available variable name if left out
`ylab="`*y label*`"`	Assigns y-axis label, uses available variable name if left out
`xlim=c(`a,b`)`	Sets horizontal axis range from a to b
`ylim=c(`c,d`)`	Sets vertical axis range from c to d

As was seen in the construction of Figure 4.8, if the argument `axes=FALSE` is included in the `plot` function, then the axes are not plotted. Moreover, using `type="n"` instructs R not to plot any points. This is a useful feature in preparing customized plots.

To demonstrate the use of the `plot` function, first clear the workspace, then recall and (for convenience) `attach` the dataset `MultiNum` from Chapter 3 using, for example,

```
MultiNum<-read.table(
   "z:/Docs/RCompanion/Chapter3/Data/Data03x04.txt",header=T)
attach(MultiNum)
```

Then, both `plot(x1,y)` and `plot(y~x1)` produce Figure 4.15.

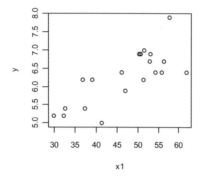

FIGURE 4.15: Basic scatterplot of y against x_1 using data from dataset `MultiNum`.

Including `type="l"` in the `plot` function call will produce a plot with lines drawn between consecutive points. This does not always result in a meaningful plot. The default setting of plot type is `type="p"`.

To be on the safe side, detach `MultiNum` before continuing.

4.11   Matrix Scatterplots

The next method available to the `plot` function that will find use is `plot.data.frame`. For example, Figure 4.16 is produced using the usual `plot` function directly on the data frame `MultiNum`. In the event that the variables are not stored in a data frame, the same plot can be obtained using the `pairs` function. For example, the commands

```
plot(MultiNum)
with(MultiNum,pairs(cbind(y,x1,x2,x3)))
```

both produce Figure 4.16.

4.12   Bells and Whistles

The package `lattice` [57] provides some extremely powerful and elegant capabilities in the graphics arena. If you are likely to use a lot of graphics in

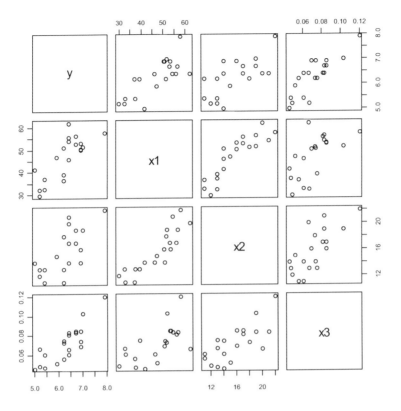

FIGURE 4.16: Matrix scatterplot of all pairs of variables in dataset `MultiNum`.

your work, this is the package to get comfortable with. If this does not apply to you, a brief summary of some "after-the-fact" functions that can be used to add features to a graphics image is shown later.

Some of these functions have already been used in earlier illustrations; others will show up in later chapters. All of the functions listed are contained in package `graphics`. In addition, see the R documentation pages for `plotmath` for help and pointers on syntax for inserting mathematical annotation in figures and more on the subject of text drawing.

In using any add-on features, it is usually best to begin with a default plot and then determine what features might be added. If you choose to change an add-on, you should generate a new plot before adding the changed feature; otherwise, the changed feature will just be slapped onto the earlier version resulting in a bit of a mess.

Function	Purpose
`title`	Inserts titles and axes labels
`axis`	Inserts an axis at a desired location with desired features
`abline(h=`*a*`)`	Plots a horizontal line $y = a$
`abline(v=`*c*`)`	Plots a vertical line $x = c$
`abline(a=`*y-intercept*`,b=`*slope*`)`	Plots the line $y = a + bx$
`lines(x,y)`	Plots lines between consecutive (x, y) pairs from `(x,y)`
`text`	Inserts text into graphics image
`legend`	Inserts a legend in the graphics image
`mtext`	Inserts text in image margins

4.13   For the Curious

Remove all objects from the workspace and then generate a random sample, say

```
e<-rnorm(35,mean=3,sd=.25)
```

The following illustrations use only basic commands seen thus far.

4.13.1   QQ normal probability plots from scratch

The goal is to plot an observed data distribution against what might be expected from the standard normal distribution. The observed quantiles (data) are first arranged in ascending order, then plotting positions are used to obtain corresponding expected quantiles. Among many, one formula used to estimate plotting positions is due to Blom,

$$p_i = \frac{i - 0.375}{n + 0.25},$$

where i is the index of the corresponding, sorted, observed quantile and n is the sample size. The final step is to plot the observed and expected quantile pairs.

```
#Obtain the sample's length
n<-length(e)
#Store the sorted observed quantiles
y.vals<-e[order(e)]
```

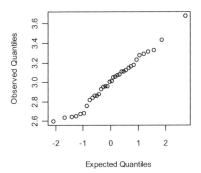

FIGURE 4.17: QQ normal probability plot from scratch using Blom's plotting positions to determine expected quantiles.

```
#Compute the plotting positions
p<-(order(y.vals)-0.375)/(n+0.25)
#Compute the theoretical quantiles
x.vals<-qnorm(p)
#Plot the observed values against the expected values
plot(y.vals~x.vals, main="QQ normal Probability Plot",
    xlab="Expected Quantiles",ylab="Observed Quantiles")
```

The result is shown in Figure 4.17, which appears to be a duplicate of the QQ plot of **e** that is obtained if the function **qqnorm** is used.

4.13.2 Half-normal plots from scratch

The process is similar to that used for the QQ plot, except that here the absolute values of the observed quantiles are used and the formula used to compute plotting positions, suggested in [45, pp. 595–596], is[3]

$$p_i = \frac{i + n - 0.125}{2n + 0.5}.$$

Basic code for a half-normal plot of the above data is shown below.

---

[3] In the function **halfnorm** from package **faraway**, the plotting position used is

$$p_i = \frac{n + i}{2n + 1}.$$

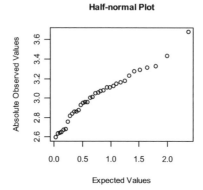

FIGURE 4.18: Half-normal plot from scratch using the approach described in [45, pp. 595–596].

```
#Obtain the sample's length
n<-length(e)
#Store the sorted absolute values of the observed quantiles
y.vals<-sort(abs(e))
#Compute the plotting positions
p<-(order(y.vals)+n-0.125)/(2*n+0.5)
#Compute the theoretical quantiles
x.vals<-qnorm(p)
#Plot the observed values against the expected values
plot(y.vals~x.vals, main="Half-normal Plot",
    xlab="Expected Values",ylab="Absolute Observed Values")
```

The result is shown in Figure 4.18. This figure differs somewhat from the figure obtained using the function `halfnorm` since a different formula was used to obtain plotting positions, and since R was permitted to choose its own `ylim` values from the sample.

4.13.3   Interpreting QQ normal probability plots

There are some typical "symptoms" in QQ normal probability plots that can be used to diagnose certain distributional characteristics in data.

Symptom	Possible Diagnosis
Concave up	Right-skewed distribution
Concave down	Left-skewed distribution
"s" shaped	Short-tailed distribution
"Backward s" shaped	Long-tailed distribution
Step-like	Rounding issues
Gaps	Missing data
Lonely points in tails	Outliers
Other strange patterns	Data obtained from mixed distributions

Recall if the plotted points in a QQ plot cluster about the qqline or the line $y = x$, as the appropriate case may be, then it is likely that the underlying random variable is approximately normally distributed. An example of the last symptom listed above can be observed by executing (figure not shown)

```
with(Salmon,qqnorm(Weight))
```

By construction, the dataset **Salmon** contains simulated samples from three different normal distributions.

Chapter 5

Automating Flow in Programs

5.1 Introduction 83
5.2 Logical Variables, Operators, and Statements 83
    5.2.1 Initializing logical variables 84
    5.2.2 Logical operators 84
    5.2.3 Relational operators 85
5.3 Conditional Statements 87
    5.3.1 If-stop 87
    5.3.2 If-then 88
    5.3.3 If-then-else 88
    5.3.4 The ifelse function 89
5.4 Loops 90
    5.4.1 For-loops 90
    5.4.2 While-loops 91
5.5 Programming Examples 91
    5.5.1 Numerical solution of an equation 92
    5.5.2 A snazzy histogram 93
5.6 Some Programming Tips 96

5.1 Introduction

This chapter completes the basic programming toolkit with brief introductions to methods used in decision making and control features in programming. Such features enable the preparation of programs that, subject to pre-defined conditions, can be used on a wider range of applications.

Before beginning this chapter, detach all objects using the `detach` command with help from `search()` and clear the workspace of all objects using the `rm` command.

The first task in preparing the way is to formalize decision-making statements.

5.2 Logical Variables, Operators, and Statements

A *logical variable* is one that can take on one of two *logical values*, `TRUE` or `FALSE`. As with numeric variables, certain (*logical*) *operations* can be performed on logical values, combinations of which result in *logical statements*. Such statements then serve to provide programs with decision-making capabilities.

5.2.1 Initializing logical variables

Logical variables can represent (logical) data structures just as was the case for previously encountered variables and can be defined by assigning a value; for example,

```
A<-FALSE; B<-TRUE
```

define the variables A and B as logical, and having the assigned values. The values TRUE and FALSE can also be entered using simply T and F. For this reason, the letters T and F should be considered reserved letters and should not be used as variable names.

Another way to initialize a logical variable is to use, for example,

```
X<-logical(1); Y<-logical(10)
```

Recall that a logical variable initialized in this manner is always assigned the default value FALSE. The variable X has a single entry, while the variable Y is a vector with 10 entries.

5.2.2 Logical operators

A *logical operator* is a function that takes, as input, one or two logical values and outputs a new value. Four basic operators find use in most programming tasks. Let A and B be logical variables, each with exactly one entry. Then, here are some basic computational examples; the variables A and B are assigned different values and the indicated operations are performed. As shown a little later in this chapter, these operators can be used to construct more complex logical statements which can be used to serve as "gate-keepers" for computational tasks (see help("&") for further details).

Variables		Operations and Output			
A	B	!A	A&B	A|B	xor(A,B)
TRUE	TRUE	FALSE	TRUE	TRUE	FALSE
TRUE	FALSE	FALSE	FALSE	TRUE	TRUE
FALSE	TRUE	TRUE	FALSE	TRUE	TRUE
FALSE	FALSE	TRUE	FALSE	FALSE	FALSE

A useful R function for evaluating the truth value (logical value) of a logical variable or statement is the function isTRUE; for example,

```
> A<-F; isTRUE(A)
[1] FALSE
```

To be useful, however, the input for this function must have exactly one entry (not a vector) and no attributes, such as a name or an index. This function will return a value of TRUE if and only if the input is TRUE. If the input has more than one entry or has a name, the logical value returned will be FALSE. For example,

```
> A<-T; isTRUE(A)
[1] TRUE
> A<-c(T,T); isTRUE(A)
[1] FALSE
> A<-T; names(A)<-"A"; isTRUE(A)
[1] FALSE
```

For those unfamiliar with the basics of *Boolean algebra*, any elementary text on logic will serve as a good resource.

5.2.3 Relational operators

The usual *relational symbols* find use in programming when decisions need to be made on how to proceed. When these symbols are used to relate two values, the result is a *logical statement*. For example,

$5 < 6$ is a TRUE statement

$5 \geq 6$ is a FALSE statement

$x_i < 5$ for all $x_i = i$, $i = 1, 2, \ldots, 10$ is a FALSE statement

In this setting, relational symbols also serve as *logical operators*. The syntax and meaning for more commonly encountered relational operators are shown below. For the sake of simplicity, let x be a vector and b a scalar (or a vector of length equal to that of x).

Operator	Syntax	Outputs TRUE for each entry of x if
==	x==b	$x_i = b$ (or if $x_i = b_i$)
!=	x!=b	$x_i \neq b$ (or if $x_i \neq b_i$)
<=	x<=b	$x_i \leq b$ (or if $x_i \leq b_i$)
<	x<b	$x_i < b$ (or if $x_i < b_i$)
>	x>b	$x_i > b$ (or if $x_i > b_i$)
>=	x>=b	$x_i \geq b$ (or if $x_i \geq b_i$)

As illustrations,

```
> x<-5; x<6
[1] TRUE
```

```
> x!=5
[1] FALSE
> isTRUE(x<=5)
[1] TRUE
```

and

```
> y<-c(1:3); y<2
[1] TRUE FALSE FALSE
> y!=2
[1] TRUE FALSE TRUE
> isTRUE(y<=3)
[1] FALSE
```

Notice that isTRUE(y<=3) compares a length-3 logical vector (i.e., y[1]<=3, y[2]<=3, y[3]<=3) against a length-1 logical vector (i.e., TRUE); hence, the output FALSE.

Consider some more involved illustrations:

```
> z<-seq(3,1,-1); y>z
[1] FALSE FALSE TRUE
> y!=z
[1] TRUE FALSE TRUE
> isTRUE((z[1]<=y[1])&(z[2]<=y[2])&(z[3]<=y[3]))
[1] FALSE
> isTRUE((z[1]<=y[1])|(z[2]<=y[2])|(z[3]<=y[3]))
[1] TRUE
```

The statements y>z and y!=z perform pairwise comparisons, while the last two statements test the value of a *single* argument obtained from a collection of logical operations.

As should be evident, it is important to determine whether the relational operator being evaluated is to be of the form $x < 5$ (a scalar versus a scalar), $y_i < 2$ for all $i = 1, 2, 3$ (a vector versus a scalar), $z_i < y_i$ for all $i = 1, 2, 3$ (two vectors of the same length), or some combination of these.

5.3 Conditional Statements

In any programming task it is helpful to be able to control the execution of certain commands or operations. For example, the following illustrates problems that can arise when computations are dependent on function domains.

```
> x<- -2:2; round(sqrt(x),3)
[1] NaN NaN 0.000 1.000 1.414
Warning message:
In sqrt(x) : NaNs produced
> asin(x)
[1] NaN -1.570796 0.000000 1.570796 NaN
Warning message:
In asin(x) : NaNs produced
```

Observe that both functions were evaluated for the given values, but in certain cases NaNs were produced as output. In certain situations, particularly if the output of a task is to be used in subsequent computations, it might be desirable either to stop a computation because certain domain issues are not satisfied or perform some alternative task. Here are a variety of options available in R.

5.3.1 If-stop

The syntax for this statement is

if (condition is TRUE) stop(Error message of why stopped)

Here is a simple example using the numbers in x from above.

```
> if (length(x[x<0])>0) stop("Negative data present")
Error: Negative data present
```

In this case, the computation is stopped and the user is alerted that negative data are present. The presence of negative x_i is tested by looking at the length of the list of negative x's.

The **if-stop** statement might be used if R is to be instructed **not** to perform a computation or **not** to continue along a particular line of tasks subject to a given condition.

5.3.2   If-then

The syntax for this statement is

> if (condition is TRUE) {Perform these tasks}

Here is a simple example using the numbers in x from above.

```
> if (length(x[x<0])==0) {sqrt(x)}
```

Nothing happens! Since the entry condition for the **if-then** statement is not satisfied, the line is skipped. On the other hand,

```
> y<-0:4; if (length(y[y<0])==0){sqrt(y)}
[1] 0.000000 1.000000 1.414214 1.732051 2.000000
```

Since there are no negative numbers in y, the instructions are to go ahead. Another approach that might find use is to give R two alternatives.

5.3.3   If-then-else

The syntax for this statement is

> if (condition) {Perform these tasks if condition is TRUE
>
>   } else {Perform these tasks if condition is FALSE}

Using x for sample numbers again, and following the same theme, running the code

```
if (length(x[x<0])==0){sqrt(x)
    } else stop("Negative data present")
```

produces

```
Error: Negative data present
```

This instructs R to calculate square roots only if there are no negative numbers in x. On the other hand, code of the form

```
if (length(x[x<0])==0){sqrt(x)
    } else {z<-x[x>=0]
        print("Negative data excluded",quote=F)
        sqrt(z)}
```

will produce

```
[1] Negative data excluded
[1] 0.000000 1.000000 1.414214
```

Here R is instructed to go ahead, regardless of the presence of negative numbers. However, the user is alerted to the presence of negative numbers and only the positive numbers are involved in the subsequent calculations.

When writing an **if-then-else** statement that extends over several lines, it is important to remember to give R something more to look for at the end of each (incomplete) command line. To ensure this happens, place the open brace, "{", on the same line after the `if` statement entry condition. The same applies to the `stop` command in the **if-stop** statement.

The `else` command should appear on the same line after the closing brace, "}", for the **if-then** portion's command line(s). The opening brace for the `else` statement should appear on the same line after the `else` command.

Nested **if-then-else** statements provide for further control in programming. Examples of such statements follow in later illustrations.

5.3.4 The ifelse function

The `ifelse` command is a (conditional) function that outputs a value subject to whether the condition value is TRUE or FALSE. Here is a very simple example:

```
> ifelse(TRUE,"Apples","Oranges")
[1] "Apples"
> ifelse(FALSE,"Apples","Oranges")
[1] "Oranges"
```

Earlier, it was mentioned that R does not like calculating roots for negative numbers, even if the root in question is a "legal" computation. For example,

```
> (-27)^(1/3)
[1] NaN
```

whereas the answer should be -3. The `ifelse` function can be used here as follows:

```
> x<-27; ifelse(x>=0,x^(1/3),-abs(x)^(1/3))
[1] 3
> x<- -27; ifelse(x>=0,x^(1/3),-abs(x)^(1/3))
[1] -3
```

A more involved application of the `ifelse` function appears later. One caution with using this function is that while it will work for vector computations, it is best to restrict activities to scalar operations or, if absolutely necessary, make sure that the test condition can have only a single value (i.e., TRUE or FALSE), as opposed to being a Boolean vector.

5.4   Loops

Loops are structures that permit repetitive tasks either by specifying the number of times a task is to be repeated or by imposing conditions that instruct when a task can be repeated. Two types of loops are presented here.

5.4.1   For-loops

Sometimes a sequence of similar computations needs to be performed a known number of times. This describes a typical situation in which a **for-loop** is useful. The general structure of a **for-loop** is

```
for (counter in counter.list) {
    Command lines}
```

The manner in which this loop is defined provides quite a bit of flexibility. The "counter" can represent a vector index in which case the "counter.list" would be the set of allowable indices. For example,

```
m<-3; b<-2; xy.pairs<-matrix(nrow=6,ncol=2)
for (i in 0:5){
    xy.pairs[i+1,]<-cbind(i,m*i+b)}
```

produces a list of ordered pairs for a line.

As another example, consider generating a geometric sequence of the form $\{a, ar, \ldots, ar^{n-1}\}$ for a given starting value, ratio, and length. The following code does this:

```
a<-3; r<-5/6; n<-25; x<-numeric(n)
for (i in 1:n){x[i]<-a*r^(i-1)}
```

Another interesting feature of the **for-loop** is that the "counter.list" need not be numeric. For example, the loop

```
for (Data in c("Set 1","Set 2","Set 3")){
    print(paste("Use Data ",Data,sep=""),quote=F)}
```

produces

```
[1] Use Data Set 1
[1] Use Data Set 2
[1] Use Data Set 3
```

It should be observed that once initiated, a **for-loop** will always start and it stops when the *exit condition* is met; that is, the last value in the "counter.list" has been used.

5.4.2 While-loops

There are times when a sequence of computations may be desired, but *only if* a certain *entry condition* is satisfied. This describes a situation in which a **while-loop** is useful. A **while-loop** uses a logical entry condition that can change in value at any time during the execution of the loop. The general structure of such loops is

```
while (entry condition) {
        command lines if the entry condition is TRUE}
```

There are many situations when a **while-loop** is more appropriate than a **for-loop**. Consider finding the length of a data list, say x (pretend that the length function does not exist). First, generate a list of unknown length[1]

```
x<-rbinom(rbinom(1,250,.5),150,.5)
```

#Note: The inner rbinom randomly picks a sample size

#Initialize a length counter

```
l<-0
```

#Add one to length counter if next entry is available

```
while (paste(x[l+1])!=paste(NA)) {l<-l+1}
```

#Output the result

```
print(paste("Length of x is",l),quote=F)
```

A point to note here concerns the specific entry condition for this illustration. The mode of each entry in x is numeric (if not empty), whereas the mode of NA is logical. To enable a meaningful comparison (test), the mode of both quantities, the entry of x and the "value" NA needs to be coerced into the same type. The paste function performs this task by converting both "values" into character mode.

5.5 Programming Examples

Two examples of preparing fairly basic programs using R syntax are presented here. The first illustrates the use of a while-loop and the ifelse function, and the second provides an illustration of how one might prepare a function to perform a particular task given any univariate numeric dataset.

---

[1]The function rbinom generates binomially distributed random samples. See help(rbinom) for more.

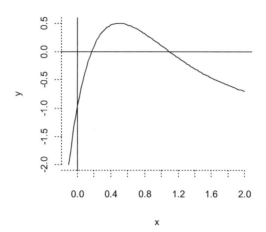

FIGURE 5.1: Graph of $y = 3xe^{1-2x} - 1$ showing lower of two x-intercepts lies within the interval $(0, 0.5)$.

5.5.1   Numerical solution of an equation

A more technical application of the **while-loop** might be to find the smaller solution of the equation $3xe^{1-2x} - 1 = 0$ using a very basic numerical algorithm. From a plot of $y = 3xe^{1-2x} - 1$ (see Figure 5.1), it is evident that two solutions exist, the smaller being in the interval $(0, 0.5)$.

Consider using the midpoint algorithm to approximate the smaller of the two x-intercepts. First, define the function $f(x) = 3xe^{1-2x} - 1$,

```
f<-function(x){return(3*x*exp(1-2*x)-1)}
```

Starting with the interval $(0, 0.5)$, the midpoint algorithm first determines which half of this interval contains the x-intercept. In this case it is the lower half, $(0, 0.25)$. The process is then repeated until the value of f at the midpoint is suitably close to zero. So,

> #Initialize the left and right points of the starting interval
>
> a<-0;b<-.5
>
> #Initialize a loop counter
>
> i<-integer(1)
>
> # Start the midpoint algorithm
>
> #Enter loop if f at midpoint is significant to 5 decimal places and

#as long as the number of iterations is less than 1000

```
while ((abs(round(f((a+b)/2),5))>=.00001)&(i<=1000)){
        #Pick the half-interval that contains the x-intercept
        if (f(a)*f((a+b)/2)>0){a<-(a+b)/2} else {b<-(a+b)/2}
        #Update the loop counter and signal end of while loop
        i<-i+1}
#Output the result
print(paste("Solution: x = ",round((a+b)/2,3),
        "; Iterations = ",i,sep=""),quote=F)
```

Notice the inclusion of comment lines, indicated by the # symbol, and the use of the paste function. If this code is run, the result is

```
[1] Solution: x = 0.173; Iterations = 17
```

So, after 17 iterations, and to a tolerance of 0.00001, the approximate solution rounded to three decimal places is $x = 0.173$.

5.5.2   A snazzy histogram

Now, consider obtaining a jazzed up density histogram for any given data list, say

```
x<-rnorm(50,mean=23,sd=5)
```

It would be useful to list the features desired in any such density histogram. This helps in preparing the code.

- Use a custom sized window with suitable text and scaling.

- Superimpose a normal density curve on the histogram.

- Customize the horizontal and vertical axes limits.

- Customize the tick marks on the axes.

- Label axes appropriately.

- Put the mean, median, and standard deviation of the data on the figure.

- Place the code in a function, call this function snazzy.hist.

Address these features one by one. Open a new script file and then create a histogram object that contains all necessary information, except the plot:

```
win.graph(width=3,height=3);par(ps=10,cex=.75)
chart<-hist(x,plot=FALSE)
```

Included in the `names` of objects in the object `chart` is `breaks`, which contains the break points (class boundaries) of the histogram produced. These break points are now used to develop the density curve and enhance the tick mark displays on the axes.

First, `attach` the object `chart` for easy access to its contents. Then obtain the number of bins and class width.

```
attach(chart)
#Obtain number of bins and class width
bins<-length(breaks)
bin.width<-breaks[2]-breaks[1]
```

Next, to develop a suitable range of x-values and suitable density curve points, start with the point that is one class width below the lowest break point and end with the point that is one class width above the last break point. Divide the range of *x*-values by 100 to get the increment for consecutive *x*-values.

```
xmin<-breaks[1]-bin.width; xmax<-breaks[bins]+bin.width
xinc<-(xmax-xmin)/100
xv<-seq(xmin,xmax,xinc)
```

Now, generate the corresponding normal densities using the sample mean and standard deviation

```
yv<-dnorm(xv,mean=mean(x),sd=sd(x))
```

To customize the horizontal axis tick marks exactly at class boundaries, store locations of horizontal axis tick marks.

```
xticks<-seq(xmin,xmax,bin.width)
```

Now, using nested `ifelse` functions and the object `density` in `chart`, determine a suitable number of decimal places to use in customizing the increments for the vertical axis tick marks, then store the locations.

```
#If max(density)>0.5 use 1 dec. place
ydec<-ifelse(max(density)>0.5,1,
    #Else if 0.05<max(density)<=0.5 use 2 dec. places, else use 3
    ifelse((max(density)>0.05)&(max(density)<=0.5),2,3))
#Get the y-increment
yinc<-round(max(density)/5,ydec)
#Get the maximum y-value
ymax<-round(max(density),ydec)+yinc
#Obtain and store the tick mark locations
yticks<-seq(0,ymax,yinc)
```

Plot the `chart` object as a density histogram with no axes or labels using the ranges of x- and y-values obtained above.

```
plot(chart,freq=FALSE,axes=FALSE,main=NULL,xlab=NULL,
     ylab=NULL,xlim=c(xmin,xmax),ylim=c(0,ymax))
```

Now include the axes with tick marks at locations determined above as well as relevant titles and labels.

```
axis(1,at=xticks); axis(2,at=yticks)
title(main=paste("Histogram and Normal Curve for",
#Use the deparse function to extract the variable name
     deparse(substitute(x))), xlab=deparse(substitute(x)),
                    ylab="Probability Density")
```

Include the probability density curve, sample mean, and standard deviation at specified locations, and finally detach the chart object.

```
lines(xv,yv,lty=3)
mtext(bquote(bar(.(substitute(x)))==.(round(mean(x),2))),
                    at=max(xv),line=0,adj=1)
mtext(bquote(s==.(round(sd(x),2))),
                    at=max(xv),line=-1,adj=1)
mtext(bquote(med==.(round(median(x),2))),
                    at=max(xv),line=-2,adj=1)
detach(chart)
```

Once a program works (see Figure 5.2), and if it appears to be one that might find repeated use, it is useful to use it to build and save as a function to be called up whenever needed. It is also helpful to set aside some simple and clear instructions on the types of arguments that are passed into the function along with a brief description of what is produced by the function.

To prepare the function, insert the code outlined above within the function definition shown below.

```
snazzy.hist<-function(x){
        win.graph(width=4,height=4)
        par(mfrow=c(1,1),ps=10,cex=.75)
        chart<-hist(x,plot=FALSE)
        attach(chart)
```
put remaining code developed above here
```
        detach(chart)}
```

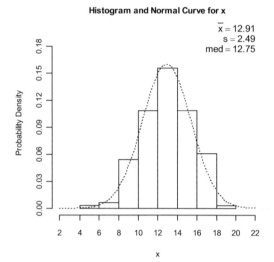

FIGURE 5.2: Density histogram with a superimposed density curve obtained using the function snazzy.hist.

and then run the code. Now, for a quick test, run

```
z<-rnorm(157,mean=13,sd=2.5)
snazzy.hist(z)
```

Once the function is working satisfactorily, it can be "dumped" in a function folder using a command of the form

```
dump("snazzy.hist",
     "Z:/Docs/RCompanion/Chapter5/Functions/snazzyhist.R")
```

This function then can be "sourced" whenever needed. Note the use of the functions deparse, substitute, bquote, . (), mtext, and paste in enhancing the graphics image. Except for mtext, which is in package graphics, all the others are in package base. Use the help function to find out more.

5.6   Some Programming Tips

Some common pointers offered to newcomers in programming include the following:

- **First understand the problem, and the goal:** Make sure you know what you are trying to achieve. Know the relevance of what is given and know what the end result should look like (at least approximately).

- **Outline the task by hand, prepare a flow-chart:** Break down the complete task into a collection of modules (sub-tasks), as much as possible; prepare a flow chart of what path the calculations (including decision points and loops) should follow in a logical fashion.

- **Trace each module using simple examples:** Use simple examples to check if each module works properly; pay close attention to if-statements and loops. You can use "echo commands" (print commands) at important points to help in this.

- **Check for compatibility between modules:** Make sure the information being passed from one module to another is compatible.

- **Style and documentation:** Wherever relevant, place a comment line in the program. In R, any line preceeded by a # is ignored by the R compiler and treated as a comment line. Use indentation to identify the body of any loop, if-statement, or continuation of a multi-line command.

- **Use a code editor, rather than a basic text editor:** This makes debugging much easier as it usually contains a "syntax spell-check" feature that alerts the programmer to syntax errors. Logical errors still remain the responsibility of the programmer. A free code editor, called **Tinn-R**, is available at http://www.sciviews.org/Tinn-R. The Mac script editor appears to have this feature.

Part II

Linear Regression Models

Chapter 6

Simple Linear Regression

6.1   Introduction 101
6.2   Exploratory Data Analysis 102
6.3   Model Construction and Fit 104
6.4   Diagnostics 106
    6.4.1   The constant variance assumption 107
        6.4.1.1   The F-test for two population variances 108
        6.4.1.2   The Brown–Forsyth test 109
    6.4.2   The normality assumption 110
        6.4.2.1   QQ normal probability correlation coefficient test 111
        6.4.2.2   Shapiro–Wilk test 112
    6.4.3   The independence assumption 112
    6.4.4   The presence and influence of outliers 113
6.5   Estimating Regression Parameters 116
    6.5.1   One-at-a-time intervals 116
    6.5.2   Simultaneous intervals 118
6.6   Confidence Intervals for the Mean Response 119
    6.6.1   One-at-a-time t-intervals 119
    6.6.2   Simultaneous t-intervals 121
    6.6.3   Simultaneous F-intervals 121
    6.6.4   Confidence bands 123
6.7   Prediction Intervals for New Observations 125
    6.7.1   t-interval for a single new response 125
    6.7.2   Simultaneous t-intervals 126
    6.7.3   Simultaneous F-intervals 127
    6.7.4   Prediction bands 128
6.8   For the Curious 129
    6.8.1   Producing simulated data 129
    6.8.2   The function ci.plot 130
    6.8.3   Brown–Forsyth test revisited 130

6.1   Introduction

This chapter covers topics likely to be encountered when dealing with the construction of simple (straight line) linear regression models from data that do not exhibit any undesireable properties.

For $i = 1, 2, \ldots, n$, let y_i represent the ith observed value of a continuous *response variable* Y and x_i the corresponding value for a *continuous explanatory variable* X. Assume X is measured without error and suppose that for each x_i the corresponding observed responses y_i are prone to *random deviations* from some unknown *mean response*. For each $i = 1, 2, \ldots, n$, denote these random deviations in the observed responses by ε_i, and let β_0 and β_1 represent unknown *regression parameters*.

The general structure of a *simple linear regression model* in *algebraic form* is

$$y_i = \beta_0 + \beta_1 x_i + \varepsilon_i.$$

Moreover, it is assumed that, for each $i = 1, 2, \ldots, n$, the *error terms* ε_i have constant variances σ^2, are independent, and are identically and normally distributed with $\varepsilon_i \sim N(0, \sigma^2)$.

In matrix form, the model has the appearance

$$\mathbf{y} = \mathbf{X}\boldsymbol{\beta} + \boldsymbol{\varepsilon},$$

where it is assumed that the *design matrix*, $\mathbf{X}$, has *full column rank*; the *response vector* $\mathbf{y}$ is a solution of $\mathbf{y} = \mathbf{X}\boldsymbol{\beta} + \boldsymbol{\varepsilon}$, and the entries of the *error vector* $\boldsymbol{\varepsilon}$ satisfy the above-mentioned assumptions.

A traditional application of simple linear regression typically involves a study in which the continuous response variable is, in theory, assumed to be linearly related to a continuous explanatory variable, and for which the data provide evidence in support of this structural requirement as well as for all fundamental assumptions on the error terms.

TABLE 6.1: Data for Chapter 6 Illustrations

x	0.00	0.75	1.50	2.25	3.00	3.75	4.50	5.25	6.00
y	54.3	50.8	58.0	54.6	45.3	47.0	51.7	43.3	44.7

x	6.75	7.50	8.25	9.00	9.75	10.50	11.25	12.00	12.75
y	38.5	42.1	40.0	32.0	34.6	32.8	33.4	28.7	26.9

Data for illustrations to follow are given in Table 6.1. This is simulated data (see Section 6.8) and is stored in the file `Data06x01.R`. The file can be loaded using the `source` function; for example,

```
source("Z:/Docs/RCompanion/Chapter6/Data/Data06x01.R")
```

The data frame is called `SimpleRegData`.

6.2 Exploratory Data Analysis

A histogram can be plotted using the function `hist`, or the earlier written function `snazzy.hist` function can be used here to obtain a density histogram with normal density curve (see Figure 6.1).

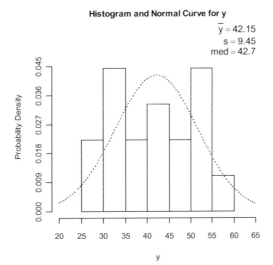

FIGURE 6.1: Density histogram of y with superimposed normal density curve using the function `snazzy.hist`.

```
source("Z:/Docs/RCompanion/Chapter5/
                      Functions/snazzyhist.R")
with(SimpleRegData,snazzy.hist(y))
```

Such figures might give some feel for clear deviations from symmetry in the observed responses.

A preliminary assessment of the manner in which the response might be (at least approximately) related to the explanatory variable can be made using a basic scatterplot[1] (see Figure 6.2).

```
with(SimpleRegData,plot(y~x,xlab="x",ylab="y"))
```

At this point, boxplots can also be used in a preliminary check for potential outlying data as well as

```
> summary(SimpleRegData)
       y                 x
 Min.   :26.90    Min.   : 0.000
 1st Qu.:33.70    1st Qu.: 3.188
```

---

[1] Note, the function `win.graph` should be used if a custom sized graphics window is desired.

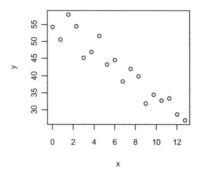

FIGURE 6.2: Scatterplot of y against x to examine the possible nature of the relationship between the two variables.

```
Median :42.70    Median : 6.375
Mean   :42.15    Mean   : 6.375
3rd Qu.:49.85    3rd Qu.: 9.562
Max.   :58.00    Max.   :12.750
```

Combining the observations that the response does not suggest a deviation from symmetry, and the fact that there appears to be an approximate linear relationship between X and Y, it is reasonable to fit the data to a simple linear regression model.

6.3   Model Construction and Fit

The basic function call for fitting the data to a simple linear regression model has the appearance

```
Simple.mod<-lm(y~x,SimpleRegData)
```

Then `names(Simple.mod)` shows that the resulting object, `Simple.mod`, contains several sub-objects each of which contains useful information. To determine what the *fitted model* is, execute

```
> round(Simple.mod$coefficients,3)
(Intercept)        x
     56.418    -2.238
```

Thus, the fitted model is $\hat{y} = 56.418 - 2.238\,x$.

Among the objects contained in `Simple.mod` that will find use in the diagnostic phase are `coefficients`, which contains the *parameter estimates* $\hat{\beta}_0$ and $\hat{\beta}_1$; `fitted.values`, which contains the *fitted values* $\hat{y}_i$; `residuals`, which contains the *residuals*, $\hat{\varepsilon}_i$, for the fitted model; and `df.residual`, the residual degrees of freedom.

In the case of simple linear regression models, it is a simple matter to obtain the traditional ANOVA table:

```
> anova(Simple.mod)

Analysis of Variance Table

Response: y
            Df   Sum Sq   Mean Sq   F value    Pr(>F)
x            1   1365.1   1365.07   144.17    2.04e-09 ***
Residuals   16   151.5      9.47
```

This table contains the *residual sum of squares* ($SSE = 151.5$) and *regression sum of squares* ($SSR = 1365.1$). The *sum of squares of total variation* ($SSTo$) is the sum of SSE and SSR. The *mean square error* ($MSE = 9.47$) is also contained in this table along with the F- and p-value for the corresponding *goodness-of-fit* F-test for the model.

It is important to remember that this approach of obtaining the traditional ANOVA table *will not work* for later multiple linear regression models.

The `summary` function can be used to obtain the model *summary statistics*, which include measures of fit. This summary also provides the essentials of the information given in the traditional ANOVA table. The summary statistics produced can be stored in an object for later recall using the command (output not shown)

```
(Simple.sum<-summary(Simple.mod))
```

Observe that by enclosing the whole assignment statement within parentheses, R not only performs the assignment, but also outputs the summary statistics. Here is a partial list of the summary output for the linear model object `Simple.mod`.

The object `Simple.sum` contains summary statistics associated with the residuals and the coefficients along with s, the *residual standard error*, stored as `sigma`; the *coefficient of determination*, r^2, stored as `Multiple R-squared`; and the relevant F-statistic and degrees of freedom for the goodness of fit F-test, stored as `fstatistic`.

Testing the model's significance via the hypotheses

$$H_0 : y_i = \beta_0 + \varepsilon_i \qquad \{\text{Reduced model}\}, \quad \text{vs.}$$
$$H_1 : y_i = \beta_0 + \beta_1 x_i + \varepsilon_i \quad \{\text{Full model}\}.$$

then boils down to interpreting the line

```
Response: y
             Df    Sum Sq   Mean Sq  F value      Pr(>F)
x             1    1365.1   1365.07  144.17    2.04e-09 ***
```

in the earlier ANOVA table, or the equivalent line

```
F-statistic: 144.2 on 1 and 16 DF, p-value: 2.04e-09
```

or the row corresponding to x in

```
Coefficients:
               Estimate   Std. Error   t value  Pr(>|t|)
(Intercept)    56.4175       1.3921      40.53  < 2e-16 ***
x              -2.2380       0.1864     -12.01 2.04e-09 ***
```

in the object `Simple.sum` generated above. Finally, the line

```
Multiple R-squared: 0.9001, Adjusted R-squared: 0.8939
```

in `Simple.sum` contains the coefficient of determination, $r^2 \approx 0.9001$. Recall, by definition, this indicates that close to 90% of the variations in the observed responses for the given data are well explained by the fitted model and variations in the explanatory variable.

Note that in simple linear regression cases, `Multiple R-squared` should be read simply as `r-Squared` and `Adjusted R-squared` should be ignored.

6.4 Diagnostics

For many, graphical diagnostic methods are considered adequate and for pretty much all, these methods are essential in providing visual support when numerical assessment methods are used. To facilitate simpler code, begin by extracting the needed information from the objects `Simple.mod` and `Simple.sum`.

```
e<-residuals(Simple.mod)
y.hat<-fitted.values(Simple.mod)
s<-Simple.sum$sigma
r<-e/s
d<-rstudent(Simple.mod)
```

The objects `e`, `y.hat`, and `s` are fairly self-explanatory, `r` contains *standardized residuals*, and `d` contains what are referred to as *studentized deleted residuals*. More on these is given in Chapter 8.

6.4.1 The constant variance assumption

To assess whether the variances of the error terms *might not be constant*, the two traditional plots used are given in Figure 6.3, the code[2] for which is

```
plot(e~y.hat,xlab="Fitted Values",ylab="Residuals")
abline(h=0)
plot(e~SimpleRegData$x,xlab="x",ylab="Residuals")
abline(h=0)
```

Here, ideal plots should not have any noticeable trends and should have the bulk of the plotted points randomly scattered and (approximately) horizontally and symmetrically bounded about the horizontal axis. These plots can also serve as a preliminary means of flagging potential outlying values of the response variable.

Two numerical methods are demonstrated below. In cases where the normality assumption is not in question, the *F-test for the difference in two population variances* between two normally distributed populations, seen in traditional elementary statistics courses, can be perfomed. The *Brown–Forsyth test* is an alternative, robust test that can be used in cases when the normality assumption is in question.

For both tests, the domain (x-values) is partitioned to obtain two groups of residuals, say Group A and Group B, that exhibit possible differences in spread. Preferably, the groups must be close in size.

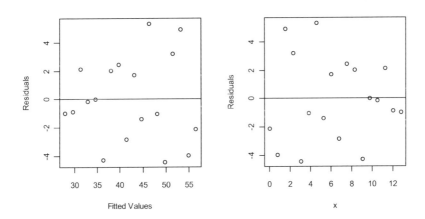

FIGURE 6.3: Plots of the residuals against fitted values and the explanatory variable for assessing the constant variance assumption.

---

[2]The graphics parameter setting `par(mfrow=c(1,2))` was used to obtain the two plots in a single window.

6.4.1.1   The F-test for two population variances

The F-test for the difference between the variances of two normally distributed populations is performed by testing the hypotheses

$$H_0 : \sigma_A^2 / \sigma_B^2 = 1 \quad \text{vs.} \quad H_1 : \sigma_A^2 / \sigma_B^2 \neq 1$$

where σ_A^2 and σ_B^2 represent the true variances for the two groups. The test statistic for this test is obtained by computing $F = s_A^2 / s_B^2$ with degrees of freedom $\nu_A = n_A - 1$ and $\nu_B = n_B - 1$. The corresponding p-value can then be obtained using the function pf introduced in Section 2.3. The tricky question to answer is usually how to partition the domain. For this, the residuals vs. x plot in Figure 6.3 can be used. Consider partitioning the domain at $x = 6$. The partitioning of the residuals can be accomplished using

```
GrpA<-residuals(Simple.mod)[SimpleRegData$x<=6]
GrpB<-residuals(Simple.mod)[SimpleRegData$x>6]
```

then

```
> df1<-length(GrpA)-1,df2<-length(GrpB)-1
> var(GrpA);var(GrpB)
[1] 13.5038
[1] 5.277357
```

Since $s_A^2 > s_B^2$, the p-value is calculated using a right-tailed probability.

```
> 2*pf(var(GrpA)/var(GrpB),df1,df2,lower.tail=F)
[1] 0.205476
```

Then, depending on the chosen level of significance, an inference may be drawn.

It turns out that package stats contains a function that performs this task. Using the same two groups, execute

```
var.test(GrpA,GrpB)
```

The lines of interest in the output are

```
F = 2.5588, num df = 8, denom df = 8, p-value = 0.2055
alternative hypothesis: true ratio of variances is not
equal to 1
```

If the normality assumption is in doubt, the above test becomes unreliable and the alternate Brown–Forsyth test may be used.

6.4.1.2   The Brown–Forsyth test

The *Brown–Forsyth test* is robust to violations of the normality assumption, and hence may be preferable if the constant variance assumption is to be tested on residuals for which the normality assumption is questionable. Basic R code for a function that will perform the Brown–Forsyth test for a regression model, along the lines of the procedure described in [45, pp. 116-117], is shown below. Let e represent the test data, x the sorting variable, and cutoff the desired partitioning value for x.

```
#Define function and argument list
bf.ttest<-function(e,x,cutoff){
#Get names of test data and sorting variable
dname<-deparse(substitute(e))
xname<-deparse(substitute(x))
#Partition test list into two groups and obtain group lengths
e1<-cbind(x,e)[x<=cutoff,2]; e2<-cbind(x,e)[x>cutoff,2]
n<-length(x); n1<-length(e1); n2<-length(e2)
#Obtain group medians
e1.med<-median(e1); e2.med<-median(e2)
#Obtain absolute deviations from medians
d1<-abs(e1-e1.med); d2<-abs(e2-e2.med)
#Obtain means of absolute deviations from medians
d1.ave<-mean(d1); d2.ave<-mean(d2)
#Compute the pooled variance of deviations from the medians
p.var<-(sum((d1-d1.ave)^2)+
                sum((d2-d2.ave)^2))/(n-2)
#Calculate the test statistic
tbf<-abs(d1.ave-d2.ave)/sqrt(p.var/n1+p.var/n2)
#Compute the p-value using the t-distribution
pv<-round(2*pt(-tbf,n-2),4)
#Store the results in a list
results<-list(statistic=c(t=tbf),parameters=c(df=n-2),
   p.value=pv, method=paste("Brown-Forsyth test ranked by",
     xname," and split at ",xname,"=",cutoff),
       data.name=dname,alternative=c("Variances are unequal"))
#Format results as hypothesis test class and output
class(results)<-"htest";return(results)}
```

The function `class` is useful in that it instructs R to format the list `results` in a manner suitable for outputting hypothesis test results.

Save the function created, `bf.ttest`, using the `dump` function; for example,

```
dump("bf.ttest",
     "Z:/Docs/RCompanion/Chapter6/Functions/RegBFttest.R")
```

This function can then be recalled using the `source` function when needed,

```
source("Z:/Docs/RCompanion/Chapter6/
                             Functions/regBFttest.R")
```

As for the previous F-test, consider partitioning the domain at $x = 6$. Then

```
> with(SimpleRegData,reg.bftest(e,x,6))
     Brown-Forsyth test ranked by x and split at x = 6
data: e
t = 1.8053, df = 16, p-value = 0.1723
alternative hypothesis: Variances are unequal
```

An equivalent alternate version along with a built-in function for this test is illustrated in the last section of this chapter.

6.4.2   The normality assumption

The QQ normal probability plot is a popular graphical assessment of the normality assumption; the residuals or standardized residuals may be used, each with an appropriate reference line. Figure 6.4 was obtained using

```
qqnorm(r,main="QQ Plot of Standardized Residuals")
abline(a=0,b=1,lty=3)
```

The line $y = x$ works as a reference line for standardized residuals, but not for unstandardized residuals. An alternative reference line that should be used for QQ plots of unstandardized data can be plotted using

```
qqline(r,lty=3)
```

which plots a line through the first and third quartiles.

An ideal QQ normal probability plot typically has plotted points that are randomly and closely clustered about the reference line. As outlined in Section 4.13.3, QQ plots can provide a fair bit of information with respect to the distribution of the error terms as well as about the observed data. Outliers in the data appear in one or both of the tails of the plot and generally tend to foul up the "niceness" of a desired linear trend in the plot.

Two numerical tests of normality are given here, for one of which R has a built-in function. The two tests are close in power; however, according to the literature, the *normal QQ correlation coefficient test* appears to be more versatile with respect to sample size than the *Shapiro–Wilk test*.

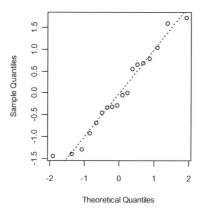

FIGURE 6.4: QQ normal probability plot with superimposed line $y = x$ for reference.

6.4.2.1 QQ normal probability correlation coefficient test

This test follows directly from the QQ normal probability plot and involves the lists used in constructing the QQ plot. Alter the earlier QQ normal probability plotting command slightly,

```
qqplot<-qqnorm(e,plot=FALSE)
```

The names of the objects contained in `qqplot` are `x`, the expected (theoretical) quantiles, and `y`, the observed (sample) quantiles. To perform the correlation coefficient test, first compute the correlation coefficient for `x` and `y`,

```
> with(qqplot,cor(x,y))
[1] 0.983763
```

then, the test value, $r_Q \approx 0.9838$, is compared with a critical value obtained from Table 2 on page 78 of Looney and Gulledge's paper.[3] This table provides a list of empirical percentage points. For this Companion, these percentage points have been placed in the function `qq.cortest` and saved in the source code file `qqcortest.R`. The call for this function is

```
qq.cortest(e,a)
```

---

[3] *Source*: Table 2, p. 78 from *Use of the Correlation Coefficient with Normal Probability Plots*, by Looney, S. W. & Gulledge, Jr., T. R., Vol. 39, No. 1 (Feb 1985), pp. 75-79. Reprinted with permission from *The American Statistician*. Copyright 1985 by the American Statistical Association. All rights reserved.

where e is the sample to be tested for normality and a is the level of significance desired. To use this function here, first load the function using, for example,

```
source("z:/Docs/RCompanion/Chapter6/Functions/qqcortest.R")
```

then the test can be performed at $\alpha = .05$ on the earlier computed residuals, e, using

```
> qq.cortest(e,0.05)

    QQ Normal Probability Corr. Coeff. Test, alpha = 0.05

data: e, n = 18

RQ = 0.9838, RCrit = 0.946

alternative hypothesis: If RCrit > RQ, Normality assumption
is invalid
```

So, $r_{\text{Crit}} = 0.946$ and since $r_Q > r_{\text{Crit}}$, the null hypothesis that the error terms are normally distributed is not rejected.

6.4.2.2 Shapiro–Wilk test

This test is a modification of the correlation coefficient test, and the associated R command to perform this test is

```
> shapiro.test(e)

        Shapiro-Wilk normality test

data: e

W = 0.9567, p-value = 0.54
```

which also indicates that there is not enough evidence to reject the assumption of normality (for $\alpha < p$).

6.4.3   The independence assumption

This assumption can be a hard one to assess; however, if the data are collected in a manner that permits any form of sequence plot of the residuals, then a sequence plot may be used to diagnose the presence of some form of dependence, either on preceding observations or indirectly through some missing variable. Thus, plots to assess the independence assumption might include, as appropriate, a time-sequence plot, a plot of the pairs $(\hat{\varepsilon}_i, \hat{\varepsilon}_{i+1})$, or plots of the residuals against some other variables that are not included in the model.

For example, trends in a scatterplot of the residuals against a spatial variable might help identify the presence of some form of *spatial auto-correlation*. Trends in a scatterplot of the residuals against time would serve to diagnose dependence on time, and a time-sequence plot (if possible for the data in question) can sometimes help diagnose the possible presence of auto-correlation.

In a time-sequence plot of the residuals, the presence of *positive first-order autocorrelation* is indicated by alternating runs of positive and negative residuals, whereas alternating positive and negative residuals indicate the presence of *negative first order autocorrelation.*

In the case of time-series data, the Durbin–Watson test can be used to test for the presence of first order autocorrelation. A function for this test is contained in the package `lmtest` [71]. Once package `lmtest` is loaded, using the function `library`, the following basic command is used

```
dwtest(model,alternative="choice")
```

where `model` represents the model in question and `choice` is one of `greater` (right-tailed test for positive autocorrelation), `two` (two-tailed test for both types of autocorrelation), or `less` (left-tailed test for negative autocorrelation). Keep in mind that the Durbin–Watson test for autocorrelation tests only for the presence of first order autocorrelation. If this test is used, be sure to `detach` the package `lmtest` and its companion package when no longer needed.

6.4.4   The presence and influence of outliers

In the case of simple linear regression, an assessment of the influence of an outlying case, an ordered pair (x_k, y_k) for which either x_k, y_k, or both x_k and y_k influence the fit of the model, is best performed using the earlier constructed xy-scatterplot in Figure 6.2. Typically, an influential outlier is indicated by a plotted point on an xy-scatterplot that lies far from the bulk of the data and that does not fit in, or appears to disrupt the general trend of the bulk of the data.

It might also correctly be inferred that if an observed response is an outlying case, then the corresponding residual will lie further from zero than the bulk of the residuals. Thus, if a particular value for the observed responses is an outlier, then this will be evident in plots shown in Figures 6.3 and 6.4.

One way to test for outliers might be to decide beforehand how many points to test. Suppose the 5% of residuals that are most extreme fit the bill. For the current data, this amounts to a single data value. Cut-off values can be obtained using a *Bonferroni adjustment* and then be placed in a plot of the studentized deleted residuals such as in Figure 6.5.

For example, suppose it is determined that the most extreme 5% of the data involve m observations, which may or may not be potential outliers. Then, cutoff values for possible outliers are obtained using $t(\alpha/(2m), n-p-2)$, where n is the sample size and, in the case of simple regression, $p = 1$. Thus, for the current model, the code

```
n<-length(e); m<-ceiling(.05*n); p<-1
```

```
cv<-qt(0.05/(2*m),n-p-2,lower.tail=F)
plot(d~y.hat,ylim=c(-cv-.5,cv+.5),xlab="Fitted Values",
        ylab="Studentized Deleted Residuals")
abline(h=c(-cv,0,cv),lty=c(3,1,3))
```

produces Figure 6.5.

While none of the plotted points for the example in question lie outside the plotted cutoff lines, there are occasions when outliers will be present. For this reason it is useful to bring up the `identify` function at this point. This function is located in package **graphics**, and the general syntax is

```
identify(x.list,y.list)
```

where $x.list$ represents the variable for the horizontal axis and $y.list$ represents the variable for the vertical axis. On execution of the `identify` command, the R graphics window becomes interactive, and moving the cursor onto the graphics window will result in the cursor taking on the appearance of cross-hairs. To label a point, move the cross-hairs as close to a point of interest and left-click the mouse. On doing this, the case number of the point in question appears on the graphics image right next to the point of interest. End the identification process by right-clicking the mouse and selecting **Stop** from the menu that pops up.[4]

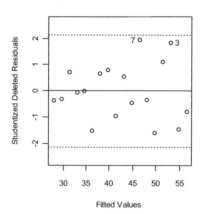

FIGURE 6.5: Plot of studentized deleted residuals against fitted values with reference lines obtained from Bonferroni cutoff values for extreme 5% of plotted points.

---

[4]Be aware that in some cases the `identify` function becomes picky and refuses to coorporate. This seems to occur when the data for the image in question do not have clear x and y coordinates, or when plotted points are too close together.

To illustrate the use of the function `identify`, suppose the two extreme points in Figure 6.5 were of interest. Then, execute

```
identify(y.hat,d)
```

and use the cross-hairs to identify the two points as described above, being sure to stop the process when done. The result will be as shown in Figure 6.5.

Package `car` [26] contains a built-in function to perform Bonferroni's outlier test and report up to a specified number of outliers (default of 10) that lie below a given joint significance level. The basic command has the form

```
outlierTest(model)
```

where *model* is replaced by the name of the model object in question. This function computes and reports both the unadjusted and the Bonferroni adjusted p-value for the absolute value of the most extreme studentized deleted residual. For example, after loading package `car`, execute

```
> outlierTest(Simple.mod)

No Studentized residuals with Bonferonni p < 0.05

Largest |rstudent|:

  rstudent  unadjusted p-value  Bonferonni p

7 1.955192            0.069455            NA
```

It is helpful to go through what is displayed here. The output indicates that observation 7 (see Figure 6.5) has the most extreme studentized deleted residual having absolute value `rstudent`

```
> max(abs(d))

[1] 1.955192
```

The unadjusted p-value for this studentized deleted residual is given under `unadjusted p-value`

```
> 2*pt(max(abs(d)),n-p-2,lower.tail=F)

[1] 0.06945538
```

and the adjusted p-value, `Bonferonni p`, which is obtained by multiplying the unadjusted p-value by $n = 18$, ends up being greater than one.

```
> 2*pt(max(abs(d)),n-p-2,lower.tail=F)*n

[1] 1.250197
```

This is indicated by `Bonferroni p` being `NA` in the output by `outlierTest`. Note that if the observation with the most extreme studentized deleted residual is not flagged as an outlier via the unadjusted and adjusted p-values, it may be concluded that outliers are not present in the observed responses.

Be sure to detach package `car` and its required packages before continuing. It is important to remember that, as executed above, the function `outlierTest` outputs results for $m = 1$ and $m = n$ for the Bonferroni adjustment. The first case may be restrictive, and the second may be considered too "forgiving" in its assessment of the outlier status of observations, particularly if the sample in question is large.

In conclusion, the use of computed Bonferroni cutoff values in a studentized deleted residual plot, such as Figure 6.5, might be preferred if a predetermined percentage of extreme cases is to be tested. The function `outlierTest` works well to test the most extreme of all studentized deleted residuals.

The influence of outliers (or any observed response for that matter) can be further assessed using a variety of influence measures; however, in the case of simple linear regression these measures are really not needed since Figure 6.2 actually provides for a very clear visual assessment of the potential influence of any flagged outlier. If a flagged outlier appears to buck the general trend of the bulk of the plotted points, this outlier may be considered influential.

6.5   Estimating Regression Parameters

Two types of interval estimates can be obtained: If only one of the two parameters needs to be estimated, a *one-at-a-time interval* is obtained; for both parameters, *simultaneous intervals* are obtained.

6.5.1   One-at-a-time intervals

In simple linear regression, an interval of this type has the form, for $j = 0$ or $j = 1$,

$$\hat{\beta}_j - t(\alpha/2, n - p - 1)\, s_{\hat{\beta}_j} \leq \beta_j \leq \hat{\beta}_j + t(\alpha/2, n - p - 1)\, s_{\hat{\beta}_j}$$

where n is the sample size, $p = 1$ is the number of explanatory variables in the model, $\hat{\beta}_j$ represents the jth parameter estimate, and $s_{\hat{\beta}_j}$ is the standard error for $\hat{\beta}_j$. The matter of computing these intervals is simple once the necessary statistics are known.

Recall that in the object `Simple.sum`, the sub-object `coefficients` contains the relevant information on the parameter estimates. That is,

```
> Simple.sum$coefficients
```

	Estimate	Std. Error	t value	Pr(>|t|)
(Intercept)	56.4175	1.3921	40.53	< 2e-16
x	-2.2380	0.1864	-12.01	2.04e-09

So, for example, the 95% confidence interval for β_1 can be obtained using the basic code:

```
#For convenience
attach(Simple.sum)
#Set level of significance
alpha<-0.05
#Extract parameter estimate and standard error
b1<-coefficients[2,1]; sb1<-coefficients[2,2]
#Compute positive critical value
cv<-qt(alpha/2,n-2,lower.tail=F)
#Detach Simple.sum to be safe
detach(Simple.sum)
#Obtain bounds and print results to screen
lb<-round(b1-cv*sb1,3); rb<-round(b1+cv*sb1,3)
print(paste(lb," < beta_1 < ",rb),quote=F)
```

The result of running this code is

```
[1] -2.633 < beta_1 < -1.843
```

Note that the relevant statistics for β_0 are stored in the first (numeric) row of the object Simple.sum$coefficients, hence relevant statistics for β_j are stored in the $(j+1)^{st}$ row of Simple.sum$coefficients.

This task can also be accomplished very simply by using the built-in function confint contained in package stats. For the model in question,

```
> confint(Simple.mod,parm="x")
      2.5 %    97.5 %
x -2.633185 -1.842907
```

The argument parm represents the (one-at-a-time) parameter to be estimated; by default (if parm is absent), intervals for all parameters are produced. For specific parameter choices use "(Intercept)" for β_0 and "x" for β_1, depending on the name of the variable being used. You can also indicate your parameter choice by the position it holds in the coefficients list; thus, 1 for β_0 and 2 for β_1. The default setting for the confidence level is 95%. To change the confidence level, the argument level must be placed in the function call with the desired confidence level. For example, the command

```
confint(Simple.mod,parm=1,level=0.99)
```

will produce the 99% confidence interval for β_0.

6.5.2   Simultaneous intervals

In the case of simple regression, Bonferroni's procedure is adequate for obtaining simultaneous intervals for both parameters. The formula for two simultaneous intervals with the Bonferroni adjustment is

$$\hat{\beta}_j - t\left(\alpha/4, n - p - 1\right) s_{\hat{\beta}_j} \leq \beta_j \leq \hat{\beta}_j + t\left(\alpha/4, n - p - 1\right) s_{\hat{\beta}_j},$$

where the notation is identical to the earlier formula. Note that the joint significance level, α, is divided by 2 so that the individual significance levels are $\alpha/2$. The code for one-at-a-time intervals can be adapted to produce simultaneous intervals.

```
attach(Simple.sum)
#Set level of significance
alpha<-0.05
#Extract parameter estimates and standard error
b<-coefficients[,1]; sb<-coefficients[,2]
#Compute positive critical value, include Bonferroni adjustment
cv<-qt(alpha/4,n-2,lower.tail=F)
detach(Simple.sum)
#Obtain bounds and print results to screen
lb<-round(b-cv*sb,3); rb<-round(b+cv*sb,3)
for (j in 1:2) {print(paste(
    lb[j]," < beta_",j-1," < ",rb[j],sep=""),quote=F)}
```

The result of running this code is

```
[1] 52.975 < beta_0 < 59.86
[1] -2.699 < beta_1 < -1.777
```

The built-in function `confint` can also be used here, keeping in mind the need for a Bonferroni adjustment. The individual significance level is scaled by a factor of $1/2$, since two parameters are being estimated simultaneously. Thus,

```
> round(confint(Simple.mod,level=1-0.05/2),3)
                1.25 %   98.75 %
(Intercept)    52.975    59.860
x              -2.699    -1.777
```

Observe that the function `confint` serves the purpose of estimating regression parameters in both scenarios quite nicely.

6.6 Confidence Intervals for the Mean Response

As with intervals for regression parameters, there may be a need to obtain a confidence interval for the true mean response at a single point, or one might wish to obtain simultaneous confidence intervals for a collection of points. First, a preliminary discussion of what is involved in the computations.

Let $\mathbf{x}'_0 = (1, x_0)$, where x_0 denotes a given value for the explanatory variable, R will store $\mathbf{x}_0$ as a column vector. Then, in algebraic form, the true response at a given x_0 is

$$y_0 = \beta_0 + \beta_1 x_0 + \varepsilon,$$

the true mean response is

$$E(y_0) = \beta_0 + \beta_1 x_0,$$

the fitted value, or estimated mean response, is

$$\hat{y}_0 = \mathbf{x}'_0 \hat{\boldsymbol{\beta}} = \hat{\beta}_0 + \hat{\beta}_1 x_0,$$

and, bringing in the design matrix mentioned earlier, the estimated variance of $\hat{y}_0$ is

$$s^2_{\hat{y}_0} = \mathbf{x}'_0 (\mathbf{X}'\mathbf{X})^{-1} \mathbf{x}_0 \, s^2.$$

For the model under consideration, the needed statistics referred to above are all contained in the object, `Simple.sum`, that is: $(\mathbf{X}'\mathbf{X})^{-1}$ is contained in `cov.unscaled`; s is contained in `sigma`; and $\hat{\boldsymbol{\beta}}$ is contained in `coefficients[,1]`, the first column of `coefficients`. These three appear in each of the interval types discussed below.

6.6.1 One-at-a-time t-intervals

The confidence interval for the mean response at a single point x_0 is computed using

$$\hat{y}_0 - t(\alpha/2, n-2)\, s_{\hat{y}_0} < E(y_0) < \hat{y}_0 + t(\alpha/2, n-2)\, s_{\hat{y}_0},$$

where $t(\alpha/2, n-2)$ is computed using a command of the form

```
qt(α/2,n-2,lower.tail=FALSE)
```

Consider obtaining a 95% confidence interval for the mean response at $x_0 = 2.5$, then the code

```
#Set sample size, significance level, and point of interest
n<-length(SimpleRegData$y); a<-0.05; x0<-c(1,2.5)
```

```
#Attach the summary object for easy access
attach(Simple.sum)
#Extract needed statistics
b<-coefficients[,1]; s<-sigma; XX.Inv<-cov.unscaled
#Detach summary object
detach(Simple.sum)
#Obtain fitted value and critical value
y0<-sum(x0*b); t.val<-qt(a/2,n-2,lower.tail=F)
#Compute standard error for computed fitted value
#with the help of matrix multiplication
s.er<-s*sqrt(t(x0)%*%XX.Inv%*%x0)
#Obtain lower and upper bounds
lb<-round(y0-t.val*s.er,4); ub<-round(y0+t.val*s.er,4)
#Output the results
print(paste("The ",round(100*(1-a)),
        "% Conf. interval at x = ",x0[2],
            " is: (",lb,",",ub,")",sep=""),quote=F)
```

produces

```
The 95% Conf. interval at x = 2.5 is: (48.6525,52.9923)
```

There is a built-in R function, contained in package `stats`, which performs the same task. To use `predict`, the point of interest is first stored in a data frame, then passed into the function.

```
> x0<-data.frame(x=2.5)
> predict(Simple.mod,x0,interval="confidence",level=.95)
        fit      lwr      upr
1 50.82243 48.65253 52.99232
```

When creating the object x0, it is important to remember to name the variable in the data frame with the same name used for the explanatory variable in the fitted model. Also, the default value for `level` is .95 and, if included, the value of `level` needs to be a number between 0 and 1. In the output, `fit` represents $\hat{y}_0$.

6.6.2 Simultaneous t-intervals

Suppose confidence intervals for the mean response at m different levels (points) of X are desired. For $k = 1, 2, \ldots, m$, let the vectors $\mathbf{x}'_k = (1, x_k)$ represent the m levels (points of interest) of the explanatory variable, with each x_k being a specified value. Then,

$$\hat{y}_k = \hat{\beta}_0 + \hat{\beta}_1 x_k, \qquad s^2_{\hat{y}_k} = \mathbf{x}'_k (\mathbf{X}'\mathbf{X})^{-1} \mathbf{x}_k \, s^2,$$

and Bonferroni's procedure uses the formula

$$\hat{y}_k - t(\alpha/(2m), n-2)\, s_{\hat{y}_k} < \mathrm{E}(y_k) < \hat{y}_k + t(\alpha/(2m), n-2)\, s_{\hat{y}_k}.$$

The coding for a one-at-a-time t-interval for the mean response can be adapted to deal with this task through the inclusion of the Bonferroni adjustment and a for-loop. However, for the present, just make use of the `predict` function.

As a simple illustration, suppose simultaneous intervals are desired at the points $x_1 = 0.5$, $x_2 = 1.0$, $x_3 = 1.5$, and $x_4 = 2.0$. Then,

```
#Create input data frame, determine number of intervals
xm<-data.frame(x=c(0.5,1.0,1.5,2.0)); m<-dim(xm)[1]
#Set joint confidence level
a<-.05
#Obtain the confidence intervals
predict(Simple.mod,xm,interval="confidence",level=1-a/m)
```

Again, when creating the object `xm`, it is important to remember to name the variable in the data frame with the same name used for the explanatory variable in the fitted model. The result will be

```
       fit      lwr      upr
1 55.29852 51.60365 58.99339
2 54.17950 50.70019 57.65880
3 53.06047 49.78992 56.33103
4 51.94145 48.87144 55.01146
```

There is an alternative to the Bonferroni procedure which might be preferred if the number of simultaneous intervals desired, m, is large.

6.6.3 Simultaneous F-intervals

The *Working–Hotelling procedure* is not affected by the number of simultaneous intervals being computed and uses the formula

$$\hat{y}_k - \sqrt{2\,F\,(\alpha, 2, n-2)}\, s_{\hat{y}_k} < \mathrm{E}(y_k) < \hat{y}_k + \sqrt{2\,F\,(\alpha, 2, n-2)}\, s_{\hat{y}_k}$$

where the terms and notation are as above. Code for this computation is very similar to the earlier code for computing one-at-a-time t-intervals of the mean response; use the same points as in the previous Bonferroni procedure.

```
#Store the points of interest and note numer of intervals
#Note, here points are stored in the format (1,x_k)
xm<-cbind(rep(1,4),x=c(0.5,1.0,1.5,2.0)); m<-dim(xm)[1]
#Initialize output variables
fitted<-numeric(m); lower<-numeric(m); upper<-numeric(m)
#Note sample size and joint significance level
n<-length(SimpleRegData$y); a<-0.05
#Attach the summary object for convenience
attach(Simple.sum)
#Extract needed statistics
b<-coefficients[,1]; s<-sigma; XX.Inv<-cov.unscaled
#Detach summary object
detach(Simple.sum)
#Compute the "critical" value
cv<-sqrt(2*qf(a,2,n-2,lower.tail=F))
#Start for-loop to get results
for (k in 1:m){
        #Find the fitted values
        fitted[k]<-sum(xm[k,]*b)
        #Find the standard errors
        s.er<-s*sqrt(t(xm[k,])%*%XX.Inv%*%xm[k,])
        #Get the lower bound
        lower[k]<-round(fitted[k]-cv*s.er,4)
        #Get the upper bound and close the for-loop
        upper[k]<-round(fitted[k]+cv*s.er,4)}
#Store the results in a data frame
F.Intervals<-data.frame(fitted,lower,upper)
```

The data frame `F.Intervals` then contains

```
    fitted    lower    upper
1 55.29852 51.7576 58.8394
2 54.17950 50.8452 57.5138
```

3 53.06047 49.9262 56.1947

4 51.94145 48.9994 54.8835

In the case of simple regression, simultaneous confidence intervals can be viewed graphically.

6.6.4 Confidence bands

The Working–Hotelling procedure can be used to obtain a confidence region for the entire regression "surface" using a suitable number of x-values. In simple regression, this "surface" translates to a band in the xy-plane. The code for the earlier Working–Hotelling procedure can be altered to obtain the desired bounds on $E(y)$ corresponding to a large number of x-values. First, let $\mathbf{z}'_k = (1, z_k)$ denote the kth row of a matrix $\mathbf{Z}$ (which may or may not be the design matrix itself) with m rows. Remember when each row of $\mathbf{Z}$ is accessed individually, R views the row as a column vector, so the transpose of $\mathbf{z}_k$ has to be used in the computations. Then,

$$\hat{y}_k = \hat{\beta}_0 + \hat{\beta}_1 z_k \quad \text{and} \quad s^2_{\hat{y}_k} = \mathbf{z}'_k (\mathbf{X}'\mathbf{X})^{-1} \mathbf{z}_k\, s^2,$$

with the interval bounds being given by

$$\hat{y}_k - \sqrt{2\,F\,(\alpha, 2, n-2)}\, s_{\hat{y}_k} < E(y_k) < \hat{y}_k + \sqrt{2\,F\,(\alpha, 2, n-2)}\, s_{\hat{y}_k}.$$

Alter the earlier code for F-intervals as follows. Here, use $m = 101$.

```
#Create the matrix Z and get sample size n
attach(SimpleRegData)
a<-floor(min(x)*10)/10;b<-ceiling(max(x)*10)/10
detach(SimpleRegData)
inc<-(b-a)/100; n<-length(SimpleRegData$y)
Z<-cbind(rep(1,101),seq(a,b,inc))
#Initialize output variables and significance level
f<-numeric(101); l<-numeric(101); u<-numeric(101)
a<-0.05
#Attach the summary object for convenience
attach(Simple.sum)
#Extract needed statistics
b<-coefficients[,1]; s<-sigma; XX.Inv<-cov.unscaled
#Detach summary object
detach(Simple.sum)
```

```
#Compute the critical value
cv<-sqrt(2*qf(a,2,n-2,lower.tail=F))
#Start for-loop to get results
for (k in 1:101){
    #Get the fitted values and standard errors
    f[k]<-sum(Z[k,]*b)
    s.er<-s*sqrt(t(Z[k,])%*%XX.Inv%*%Z[k,])
    #Get the lower and upper bounds and close the for-loop
    l[k]<-f[k]-cv*s.er; u[k]<-f[k]+cv*s.er}
#Store the results in a data frame
band<-data.frame(f,l,u)
```

Execution of this code then provides the bounds needed to construct the confidence bands. The function `matplot`, from package `graphics`, enables the superimposing of more than one plot on the same axes (see Figure 6.6) and has the following generic syntax.

```
matplot(x,y,lty=c(line.type.choices),col=c(color.choices))
```

Here x is a list of x-values and y may contain two or more lists of y-values, as is the case for `band`, which contains three.

Code used to obtain Figure 6.6 from `band` is given below.

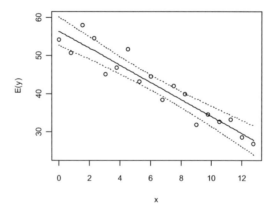

FIGURE 6.6: Confidence band of the mean response using simultaneous F-intervals obtained by the Working–Hotelling procedure.

#Plot curves for the upper and lower bounds and fitted values

```
matplot(Z[,2],band,lty=c(1,3,3),col=c(1,1,1),
              xlab="x",ylab="E(y)",type="l")
```

#Plot observed values from original data

```
with(SimpleRegData,points(x,y))
```

Note the use of the argument `col` in the function `matplot` to set all curve colors to black. By default, R uses different colors for each curve.

A built-in function for constructing confidence bands is available in package HH [39]. The use of this is illustrated in the last section of this chapter.

6.7 Prediction Intervals for New Observations

One computational difference between obtaining *confidence intervals* for the mean response and *prediction intervals* for *new responses* at a given value of X is the estimated variance used. Here, the estimated variance includes the variance of the error term corresponding to the new observation. Thus, for a prediction interval with $\mathbf{x}_0' = (1, x_0)$ being a point of interest, use

$$s_{\hat{y}_{new}}^2 = \left(1 + \mathbf{x}_0'(\mathbf{X}'\mathbf{X})^{-1}\mathbf{x}_0\right) s^2$$

in place of $s_{\hat{y}}^2$.

6.7.1 t-interval for a single new response

As with the earlier confidence intervals, a t-interval is used, the difference lying in the standard error used in the formula

$$\hat{y}_{new} - t(\alpha/2, n - 2)\, s_{\hat{y}_{new}} < y_{new} < \hat{y}_{new} + t(\alpha/2, n - 2)\, s_{\hat{y}_{new}},$$

with

$$\hat{y}_{new} = \hat{\beta}_0 + \hat{\beta}_1 x_0.$$

Replacing $s_{\hat{y}}^2$ by $s_{\hat{y}_{new}}^2$ in the earlier code for the one-at-a-time t-interval for the mean response will produce the prediction interval for y_{new} at a given point $X = x_0$.

The R function `predict` can be used to obtain prediction intervals by specifying the argument `interval="prediction"`. Consider using `predict` to obtain a prediction interval for a new response at $x_0 = 2.5$ using the default 95% confidence level.

```
> x0<-data.frame(x=2.5)
> predict(Simple.mod,x0,interval="prediction")
        fit      lwr      upr
1 50.82243 43.94783 57.69703
```

As before, remember to name the contents of the data frame x0 with the same name used for the explanatory variable in the fitted model.

6.7.2   Simultaneous t-intervals

Suppose simultaneous intervals for m new observations are desired, and

$$\mathbf{x}_1' = (1, x_1), \quad \mathbf{x}_2' = (1, x_2), \quad \ldots \quad \mathbf{x}_m' = (1, x_m).$$

Then, for $k = 1, 2, \ldots, m$, needed estimates include

$$\hat{y}_{\text{new}(k)} = \hat{\beta}_0 + \hat{\beta}_1 x_k \quad \text{and} \quad s_{\hat{y}_{\text{new}(k)}}^2 = \left(1 + \mathbf{x}_k'(\mathbf{X}'\mathbf{X})^{-1}\mathbf{x}_k\right) s^2$$

and, at a joint significance level of α, the Bonferroini prediction intervals are given by

$$\hat{y}_{\text{new}(k)} - t(\alpha/(2m), n - 2) \, s_{\hat{y}_{\text{new}(k)}}$$
$$< y_{\text{new}(k)} <$$
$$\hat{y}_{\text{new}(k)} + t(\alpha/(2m), n - 2) \, s_{\hat{y}_{\text{new}(k)}},$$

Using the R function predict, these are simple to obtain.

For Simple.mod, suppose the wish is to obtain prediction intervals for four new responses at the same four points used earlier,

$$x_1 = 0.5, \quad x_2 = 1.0, \quad x_3 = 1.5, \quad x_4 = 2.0.$$

Then

```
> xm<-data.frame(x=c(0.5,1.0,1.5,2.0))
> m<-dim(xm)[1]; a<-0.05
> predict(Simple.mod,xm,interval="prediction",level=1-a/m)
        fit      lwr      upr
1 55.29852 45.88686 64.71018
2 54.17950 44.85036 63.50864
3 53.06047 43.80716 62.31379
4 51.94145 42.75710 61.12580
```

Be sure to name the contents of the data frame xm with the same name used for the explanatory variable in the fitted model.

6.7.3 Simultaneous F-intervals

Scheffe intervals are given by

$$\hat{y}_{\text{new}(k)} - \sqrt{m\,F\,(\alpha, m, n - 2)}\,s_{\hat{y}_{\text{new}(k)}}$$
$$< y_{\text{new}(k)} <$$
$$\hat{y}_{\text{new}(k)} + \sqrt{m\,F\,(\alpha, m, n - 2)}\,s_{\hat{y}_{\text{new}(k)}}.$$

Consider obtaining Scheffe prediction intervals for the four points used in the previous illustration.

The code for the earlier Working–Hotelling procedure is altered in two spots, see included comments.

```
#Store points of interest and number of intervals desired
xm<-cbind(rep(1,4),c(0.5,1.0,1.5,2.0)); m<-dim(xm)[1]
#Initialize output variables, sample size, and significance level
fitted<-numeric(m); lower<-numeric(m); upper<-numeric(m)
n<-length(SimpleRegData$y); a<-0.05
#Attach model summary and get model statistics
attach(Simple.sum)
b<-coefficients[,1]; s<-sigma; XX.Inv<-cov.unscaled
detach(Simple.sum)
#Compute the critical value — note change from earlier code
cv<-sqrt(m*qf(a,m,n-2,lower.tail=F))
#Start for-loop to get stuff
for (k in 1:m){
        fitted[k]<-sum(xm[k,]*b)
        #Get standard errors — note change from earlier code
        s.er<-s*sqrt(1+t(xm[k,])%*%XX.Inv%*%xm[k,])
        #Get lower and upper bounds
        lower[k]<-round(fitted[k]-cv*s.er,4)
        upper[k]<-round(fitted[k]+cv*s.er,4)}
F.Intervals<-data.frame(fitted,lower,upper)
```

Then, the contents of F.Intervals can be looked at.

```
> F.Intervals
     fitted   lower   upper
1 55.29852 43.6953 66.9017
```

```
2 54.17950 42.6780 65.6810

3 53.06047 41.6525 64.4685

4 51.94145 40.6185 63.2644
```

Prediction bands can be obtained in a manner analogous to that for confidence bands.

6.7.4   Prediction bands

At times one might not know in advance what *single* point is of interest. In such cases, it might be useful to construct a prediction band *based on a collection of single point prediction intervals* (see Figure 6.7). A quick approach would be to use the function `predict` as follows.

```
X<-data.frame(x=SimpleRegData$x)

p.band<-data.frame(predict(Simple.mod,X,
                interval="prediction"))

with(SimpleRegData,matplot(x,p.band,lty=c(1,3,3),
        lwd=c(1,2,2),type="l",col=1,xlab="x",ylab="y")

with(SimpleRegData,points(x,y,pch=18))
```

Note the use of the argument `lwd` to set the line widths and `pch` to set the type of symbol to be plotted.

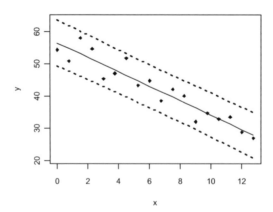

FIGURE 6.7: Prediction bands for new responses obtained using sample data *x*-values and the function `predict`.

6.8   For the Curious

As before, this section and others like it present additional topics that may or may not be needed in later sections, but might provide additional ideas in using R. Here, the idea of generating data for use in exploring code development is briefly revisited. The function `ci.plot` from package `HH` is demonstrated, and then the Brown–Forsyth test is revisited from the F-test point of view.

6.8.1   Producing simulated data

The data used for the earlier illustrations were generated using the function `rnorm`. The idea was to produce some generic, and "nice," data that could be fitted to a simple linear regression model of the form

$$
\begin{aligned}
y &= f(x) + \varepsilon \\
&= \beta_0 + \beta_1 x + \varepsilon
\end{aligned}
$$

and for which the error terms, ε, are simulated to come from a normal distribution. The data were then stored in a data frame and saved using the `dump` function. For example, the data in Table 6.1 were generated using the code

```
x<-seq(0,12.75,.75)
y<-round(55-1.9*x+rnorm(length(x),mean=0,sd=3.1),1)
#Plot to check for outrageous cases
plot(y~x)
#If happy with the data, store and save
SimpleRegData<-data.frame("y"=y,"x"=x)
dump("SimpleRegData",
        "Z:/Docs/RCompanion/Chapter6/Data/Data06x01.R")
```

The choices for values of β_0, β_1, and `sd` in the code to generate `y` were quite arbitrary. The `plot` function was used to see how the data looked before being accepted for use.

Keep in mind that it is highly unlikely that two different runs of generating `y` will produce the same "observed" responses, so this little routine can theoretically produce quite a variety of sample datasets. Note also that the nature of the simulated data can be tweaked simply by varying the standard deviation in `rnorm`, altering the function f in

$$
y = f(x) + \varepsilon,
$$

as well as by choosing a different random number generator (other than `rnorm`) for the error terms. Such tweaking can produce nice examples of diagnostic plots that simulate various violations in the error term assumptions.

6.8.2   The function ci.plot

Package HH [39] contains the function `ci.plot`, which produces and plots both confidence and prediction bands for simple regression models. The command to produce the plot (figure not shown) is

```
ci.plot(Simple.mod, conf.level=.95)
```

The default value for `conf.level` is `.95`. It is unclear which procedures are used to compute the intervals for the plotted bands in this function. Be sure to `detach` package HH and its companion packages once the task is completed.

6.8.3   Brown–Forsyth test revisited

Code for a shorter, more elegant function to perform the Brown–Forsyth test can be constructed using the approach described in [21, pp. 185–186], which makes use of the routine `lm` to perform an analysis of variance of deviations from medians. The following function duplicates the results of the earlier program, but from an F-test point of view.

```
#Define function and give argument list
bf.Ftest<-function(e,x,cutoff){
#Get test data and sorting variable names
dname<-deparse(substitute(e))
xname<-deparse(substitute(x))
#Partition list into two groups using the ifelse function
group<-ifelse(x<=cutoff,"A","B")
#Store partitioned data in a data frame
e.list<-data.frame(e,group)
#Compute group medians and deviations from medians
meds<-with(e.list,tapply(e,group,median))
dev<-with(e.list,abs(e-meds[group]))
#Fit to a linear model
info<-anova(lm(dev~group,e.list))
#Extract, store, and format results
results<-list(statistic=c(F=info[1,4]),
   parameters=c(df=info[,1]),p.value=info[1,5],
    method=paste("Brown-Forsyth test using ranking by",
     xname,"split at",xname," = ",cutoff),data.name=dname,
      alternative=c("Variances are unequal"))
class(results)<-"htest";return(results)}
```

Then,

```
with(SimpleRegData,bf.Ftest(e,x,6))
```

produces

```
Brown-Forsyth test using ranking by x split at x = 6
data: e
F = 2.0415, df1 = 1, df2 = 16, p-value = 0.1723
alternative hypothesis: Variances are unequal
```

The package `lmtest` [71] has another test for the constant variance assumption, the *Breusch-Pagan test* (for large samples), which assumes independence and normality of the error terms. Also, package `HH` has the function `hov`, which can used if the earlier grouped data, `GrpA` and `GrpB` (see Section 6.4.1.1), are made use of as follows:

```
Group<-c(rep(1,length(GrpA)),rep(2,length(GrpB)))
Group<-factor(Group)
hov(e~Group)
```

The results agree with those of `bf.Ftest` and `bf.ttest`.

Chapter 7

Simple Remedies for Simple Regression

7.1	Introduction	133
7.2	Improving Fit	134
	7.2.1 Transforming explanatory variables	134
	7.2.2 Transforming explanatory and response variables	137
7.3	Normalizing Transformations	138
	7.3.1 Transformations suggested by the theory	138
	7.3.2 Power (or log) transformations by observation	139
	7.3.3 The Box–Cox procedure	141
7.4	Variance Stabilizing Transformations	144
	7.4.1 Power transformations	144
	7.4.2 The arcSine transformation	146
7.5	Polynomial Regression	147
7.6	Piecewise Defined Models	149
	7.6.1 Subset regression	150
	7.6.2 Continuous piecewise regression	151
	7.6.3 A three-piece example	152
7.7	Introducing Categorical Variables	155
	7.7.1 Parallel straight-line models	156
	7.7.2 Non-parallel straight-line models	159
7.8	For the Curious	161
	7.8.1 The Box–Cox procedure revisited	162
	7.8.2 Inserting mathematical annotation in figures	164
	7.8.3 The split function	165

7.1 Introduction

There are two classes of transformations that may be performed: To improve fit, and to remedy violations of the normality or constant variance assumptions on the error terms. Both classes are fairly straightforward to implement; however, determining the appropriate transformation does involve a combination of a good foundation in (underlying) theoretical principles and mathematical function behavior, as well as the use of some creativity.

Aside from transformations, other methods to improve fit include the use of polynomial regression, the use of piecewise regression methods, and the inclusion of categorical explanatory variables in a model. The inclusion of additional continuous explanatory variables is discussed in the next chapter.

7.2   Improving Fit

Sometimes an xy-scatterplot of the data will suggest that y is not linearly related to x. In such cases, the fit of a model may be improved by transforming only the explanatory variable, or possibly both the response and the explanatory variable.

7.2.1   Transforming explanatory variables

Typically, if it is known that the response variable is approximately normally distributed, transformations on the response variable are avoided. In fact, one may find that a transformation of the explanatory variable will sometimes serve to improve fit as well as remedy what might seem to be violations of assumptions on the error terms.

For the following illustration, data were generated using

```
x<-runif(100,2,50); y<-2-5/x+rnorm(100,mean=0,sd=.25)
```

The data were then stored in the data frame `XTransReg` and saved to the file `Data07x01.R`. Figure 7.1 shows the plot of Y against X for this dataset. The appearance of the resulting plot does suggest the need for some thought before continuing. However, If one were to just plow ahead by fitting the data to a model of the form

$$y_i = \beta_0 + \beta_1 x_i + \varepsilon_i$$

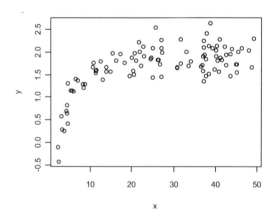

FIGURE 7.1: Scatterplot of untransformed dataset `XTransReg` showing a nonlinear relationship between y and x.

and then look at the residual plots, it would be very clear that something is not quite right. For example,[1]

```
noThink.mod<-lm(y~x,XTransReg)
with(noThink.mod,plot(fitted.values,
    residuals,xlab="Fitted Values",ylab="Residuals"))
with(noThink.mod,
    qqnorm(residuals,main="QQ Plot of Residuals"))
```

produces Figure 7.2, which confirms suspicions from Figure 7.1.

The plot of the residuals against the fitted values is a good example of a plot that suggests the possible presence of nonlinearity in the data. From the QQ normal probability plot, it might even be suspected that the error distribution is slightly skewed to the left.

In the case of simple linear regression the original scatterplot, Figure 7.1 in this case, is probably the best source of information leading to identifying a possible transformation of the data. Based on the apparent limiting behavior in the response, as X becomes large, one might suspect that a transformation of X to $1/X$ may linearize the plot. An exploratory plot using

```
with(XTransReg,plot(1/x,y))
```

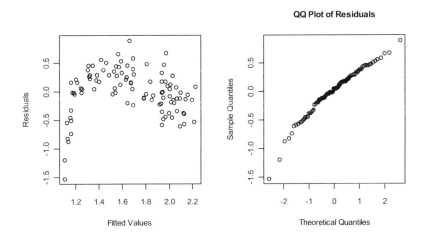

FIGURE 7.2: Plot of residuals against fitted values and QQ normal probability plot of residuals for the fitted model with untransformed data.

---

[1]The side-by-side plots in Figure 7.2 were obtained using `par(mfrow=c(1,2))`.

does, in fact, suggest an improvement, at least in the linear trend sense. So, the data might be fitted to a model of the form

$$y_i = \beta_0 + \beta_1(1/x_i) + \varepsilon_i$$

using

```
xTrans.mod<-lm(y~I(1/x),XTransReg)
```

An important point to note here is on the use of the *inhibit* function, I. Always use this function when including transformed explanatory variables in lm.

The plot of the residuals obtained from xTrans.mod shows an improvement in comparison with Figure 7.2; however, the constant variance assumption might be suspect. Figure 7.3, obtained using the following code,

```
with(XTransReg,plot(1/x,abs(residuals(xTrans.mod)),
                               ylab="|Residuals|"))
```

suggests a possible partitioning at $1/X = 0.025$. Recall the Brown–Forsyth[2] test function from the last section of Chapter 6 and execute:

```
> x<-XTransReg$x
> with(xTrans.mod,bf.Ftest(residuals,1/x,0.025))
```

The resulting output is shown below.

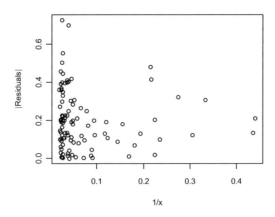

FIGURE 7.3: Plot of $|\hat{\varepsilon}|$ against $1/x$ for purposes of assessing the constant variances assumption on the error terms.

---

[2]For example, use source("Z:/Docs/RCompanion/Chapter6/Functions/RegBFFTest.R") to source the function code.

```
Brown-Forsyth test using ranking by 1/x split at 1/x =
0.025

data: residuals

F = 0.0028, df1 = 1, df2 = 98, p-value = 0.9582

alternative hypothesis: Variances are unequal
```

So, at least with the partition used, there appear to be no worries with respect to the constant variance assumption.

There will be occassions when a transformation of only the explanatory variable will not suffice.

7.2.2 Transforming explanatory and response variables

Identifying appropriate transformations for such cases is often quite difficult. For the most part, however, a scatterplot of the response against the explanatory variable combined with some knowledge of one-variable mathematical functions provides a means of determining appropriate (approximate) transformations.

For the following illustration, data were generated using

```
x<-runif(100,0,5);y<-2*x/((1+rnorm(100,mean=1,sd=.1))*x+1)
```

then stored in the data frame XYTransReg and saved to the file Data07x02.R. The plots shown in Figure 7.4 were obtained using the following code.

```
attach(XYTransReg)
plot(x,y,sub="(a)"); plot(1/x,y,sub="(b)")
plot(1/x,1/y,sub="(c)")
detach(XYTransReg)
```

Refer to these plots for the following discussion.

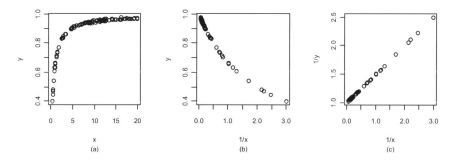

FIGURE 7.4: Using scatterplots to explore transformations that might improve fit.

Plot (a) in Figure 7.4 looks a lot like the xy-plot for the previous example, suggesting a look at the plot of Y against $1/X$ which is given in plot (b). This, in turn, suggests looking at the plot of $1/Y$ against $1/X$. Plot (c) indicates a linear trend, but with increasing spaces between plotted points. Thus, the data might be fitted to a model of the form

$$1/y_i = \beta_0 + \beta_1(1/x_i) + \varepsilon_i,$$

using

```
xyTrans.mod<-lm(1/y~I(1/x),XYTransReg)
```

Note that including the *inhibit* function I to cover the explanatory variable transformation is necessary; however, it appears that this is not necessary for the response transformation.

A quick check of the assumptions can be done using residual plots equivalent to those in Figure 7.2 for the previous illustration. Bad behavior in residual plots may occur when data are transformed, as is the case here, and there is always a possibility that outliers will be present in the transformed scale.

7.3 Normalizing Transformations

Transformations to improve fit will sometimes remedy apparent violations of error assumptions, and vice versa. Typically, a violation of the normality assumption through asymmetry indicates only the need for a transformation of the response. However, in some cases, a transformation of the response will then force the need for a transformation of the explanatory variable to improve fit.

There are a variety of approaches to choosing an appropriate transformation to remedy a violation of the normality assumption; the key is to use an approach that is theoretically sound and convenient, and then apply the (computationally) simplest transformation.

7.3.1 Transformations suggested by the theory

This approach is fairly straightforward, simply requiring an awareness of what is available in the literature. Three popular transformations include:

- If the response variable, Y, represents counts ($Y = 0, 1, 2, \ldots$), use the transformation $\sqrt{Y}$.

- If the response variable represents proportions ($0 < Y < 1$), use the transformation $\ln[Y/(1 - Y)]$.

- If the response variable represents correlations $(-1 < Y < 1)$, use the transformation $\frac{1}{2}\ln\left[(1+Y)/(1-Y)\right]$.

Once a transformation is chosen, Y is replaced by $Z = f(Y)$, its transformed scale, in the model fitting function call,

```
lm(f(y)~x,dataset)
```

The transformation for proportions is also referred to as $\text{logit}(Y)$ and the transformation for correlations is referred to as Fisher's $z(Y)$.

7.3.2 Power (or log) transformations by observation

In cases where the deviation from normality is a result of asymmetry, it is sometimes possible to achieve near normality by a trial and error approach based on observations of certain symptoms in the QQ normal probability plot of the (preferably standardized or studentized) residuals.[3]

Power transformations require that $Y > 0$ and take the form $Z = Y^\lambda$, or $Z = \ln(Y)$ if $\lambda = 0$. If a power transformation appears necessary, and the restriction $Y > 0$ is not satisfied, translate (recode) the response variable so that the translated response values satisfy the restriction.

The idea behind a power transformation is to find a λ for which the asymmetry issue is approximately corrected in the transformed scale. Typical symptoms in the QQ plot applicable to power transformations, along with suggested remedies, are:

- A QQ normal probability plot indicating a concave up trend suggests a right-skewed error distribution. Appropriate transformations include $\ln(Y)$ or Y^λ with $\lambda < 1$.

- A QQ normal probability plot indicating a concave down trend suggests a left-skewed error distribution. Appropriate transformations have the form Y^λ with $\lambda > 1$.

Once a transformation has been decided upon, fit the transformed model and recheck the QQ normal probability plot of the (preferably standardized or studentized) residuals. Repeat the process until near normality is achieved. It is suggested that the choice of λ be a "nice" number, for example, rather than choosing $\lambda = 0.46$, choose $\lambda = 0.5$. Here is an illustration.

Consider a dataset generated by

```
x<-seq(0,2,.04); y<-(x+rnorm(50,mean=2,sd=0.05))^(5/2)
```

and then stored in the data frame RightSkewed and saved to the file Data07x03.R.

---

[3] See Section 8.4 for further details on studentized residuals.

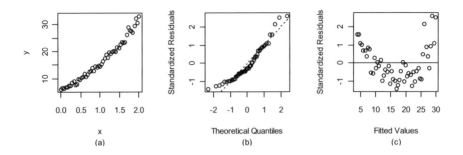

FIGURE 7.5: Diagnostic plots for the untransformed fitted model using dataset `RightSkewed`.

Fit the data to the untransformed model

$$y_i = \beta_0 + \beta_1 x_i + \varepsilon_i.$$

The plots in Figure 7.5 were obtained using the following code.[4]

```
RS.mod<-lm(y~x,RightSkewed)
r<-residuals(RS.mod)/summary(RS.mod)$sigma
f<-fitted.values(RS.mod)
with(RightSkewed,plot(y~x,sub="(a)"))
qqnorm(r,ylab="Standardized Residuals",
                main=NULL,sub="(b)")
abline(a=0,b=1,lty=3)
plot(r~f,xlab="Fitted Values",
    ylab="Standardized Residuals",sub="(c)")
abline(h=0)
```

All three plots in Figure 7.5 suggest the need for a transformation. More importantly, the QQ plot suggests the error distribution may be right skewed and the Shapiro–Wilk test (`p-value = 0.003458`) supports the suspicion that the normality assumption be rejected. Since there is no theory attached to these data to suggest an appropriate transformation, a good start might be to explore options for power transformations.

Consider using the transformation $Z = Y^{0.5}$ to shrink larger values of the response variable. Then, fitting the data to the model

$$\sqrt{y_i} = \beta_0 + \beta_1 x_i + \varepsilon_i$$

---

[4] Note that the functions `win.graph` and `par` with the argument `mfrow=c(1,3)` were used to get the three plots in the same image.

using

```
trans1.mod<-lm(sqrt(y)~x,RightSkewed)
```

and obtaining plots analogous to those in Figure 7.5 suggest *some* good things have happened. With respect to normality, the Shapiro–Wilk test appears to agree (p-value = 0.3094). This transformation is then fairly adequate; however, for the sake of argument one might, for some reason, just happen to try fitting the data to the model

$$y_i^{0.4} = \beta_0 + \beta_1 x_i + \varepsilon_i.$$

Call this `trans2.mod`. Diagnostic plots for this transformed model are shown in Figure 7.6. The Shapiro–Wilk test (p-value = 0.7593) and Figure 7.6.b suggest that pushing λ to 0.4 helped a bit with respect to normalizing the error terms. Of course, this is to be expected because of how the data were generated. It should be noted, however, even if a transformation can be pushed to improve fit or near normality, the general practice is to choose the simpler adequate transformation, in this case $Z = \sqrt{Y}$, unless the underlying theory suggests otherwise.

On a side note, the y-axes label in Figure 7.6.a was inserted using `ylab=expression(y^{0.4})` in the relevant plot command.

7.3.3 The Box–Cox procedure

This method, due to Box and Cox, provides a numerical approach to estimate an appropriate power transformation of the response for purposes of remedying a violation of the normality assumption.

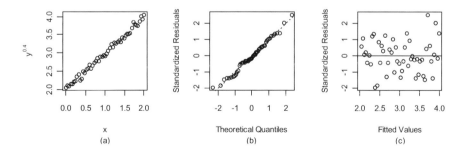

FIGURE 7.6: Diagnostic plots for the fitted model using $\hat{\lambda} = 0.4$ to transform the response in dataset `RightSkewed`.

Very briefly, the true power, λ, for the modified power transformation

$$z_\lambda = \begin{cases} \left(y^\lambda - 1\right)/\lambda & \text{if } \lambda \neq 0, \\ \\ \ln(y) & \text{if } \lambda = 0. \end{cases}$$

is estimated using a maximum likelihood method. See Section 7.8 and, for example, [42, p. 207] for more details. Once the estimate, $\hat{\lambda}$, is found, the response is transformed accordingly. Some algebra shows that the models

$$\left(y_i^{\hat{\lambda}} - 1\right)/\hat{\lambda} = \beta_0' + \beta_1' x_i + \varepsilon_i' \quad \text{and} \quad y_i^{\hat{\lambda}} = \beta_0 + \beta_1 x_i + \varepsilon_i$$

are equivalent, with

$$\beta_0 = 1 + \hat{\lambda}\beta_0'; \quad \beta_1 = \hat{\lambda}\beta_1'; \quad \text{and} \quad \varepsilon_i = \hat{\lambda}\varepsilon_i'.$$

So, in fitting the model, it is adequate to use the simpler-looking transformation. Notice, also, that if $\hat{\lambda} = 1$, then $Z_1 = Y - 1$ suggests that a power transformation should not be performed.

Functions that perform the Box–Cox procedure are contained in package MASS, named boxcox; and package car, named boxCox. The last section of this chapter provides further computational details on the Box–Cox procedure; however, for the present, the function boxcox from package MASS is demonstrated. The syntax with basic arguments for the function boxcox is

```
boxcox(model,lambda=seq(a,b,inc),eps=Tol)
```

where *model* is the original (untransformed) fitted model object; lambda, which defaults to $-2 < \lambda < 2$ with *inc* $= 0.1$, is the interval over which an estimate for λ is sought, and eps is the tolerance for setting $\lambda = 0$; this defaults to 0.02.

Using the untransformed model RS.mod, a first call of boxcox permits an inspection to enable appropriate adjustments to the horizontal scale for λ. The second call for boxcox provides a zoom into the maximum of the curve (see Figure 7.7).

The code used to obtain Figure 7.7 is:

```
library(MASS)
boxcox(RS.mod)
boxcox(RS.mod,lambda=seq(0.15,0.45,0.05))
axis(1,at=seq(0.15,0.45,.01),tcl=.25,labels=F)
```

The right-hand plot suggests using $\hat{\lambda} \approx 0.3$. Now apply this transformation both ways mentioned above:

```
lm((y^(.3)-1)/.3~x,RightSkewed)
```

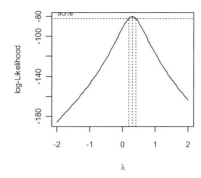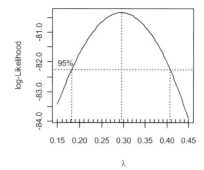

FIGURE 7.7: The Box–Cox procedure using function `boxcox` from package MASS.

to get $\hat{\beta}'_0 = 2.387$ and $\hat{\beta}'_1 = 1.860$, and

```
lm(y^(.3)~x,RightSkewed)
```

to get $\hat{\beta}_0 = 1.7161$ and $\hat{\beta}_1 = 0.5579$. Allowing for rounding, observe that

$$1 + \hat{\lambda}\,\hat{\beta}'_0 = 1 + .3 * 2.387 = 1.7161 \approx \hat{\beta}_0, \quad \text{and}$$
$$\hat{\lambda}\,\hat{\beta}'_1 = 0.3 * 1.860 = 0.558 \approx \hat{\beta}_1.$$

So, the second model fitting function call might be preferred for the sake of computational simplicity. Call this fitted model `trans3.mod`. The residual plots for this model (Figure 7.8) suggest *really* good things have happened. Compare these plots with those for the earlier transformations.

It is evident that $\hat{\lambda} \approx 0.3$ provides a better transformation to near normality than the earlier estimates of $\hat{\lambda} = 0.5$ and $\hat{\lambda} = 0.4$, and the Shapiro–Wilk test provides further evidence (`p-value = 0.9498`). However, a power transformation of this form may not make theoretical sense and it may be that a theoretically supported value for λ is close enough to an obtained estimate to be used. The confidence interval provided in Figure 7.7 provides some justification for choosing an estimate of λ that is not necessarily the data-specific maximizing value.

On a side note, observe that the function `axis` was used in the code for Figure 7.7 to include additional tick marks in the right-hand plot. These additional tick marks give a better feel for the approximate maximizing value of λ.

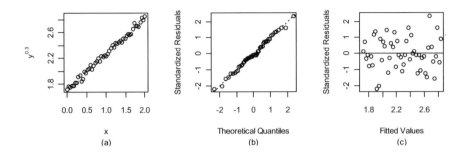

FIGURE 7.8: Diagnostic plots for fitted model obtained using $\hat{\lambda} = 0.3$ to transform the response.

7.4 Variance Stabilizing Transformations

There are occasions when a transformation of the response will also serve to stabilize the variances of the error terms as well as achieve near normality. In all cases mentioned below, the target is to find a transformation $Z = f(Y)$ such that the variances of the error terms for the transformed model

$$z_i = \beta_0 + \beta_1 x_i + \varepsilon_i$$

are, at least approximately, stabilized. Once this transformation has been found, the transformed data are fitted to a model using

```
lm(f(y)~x,dataset)
```

Be aware that it is not always possible to find a transformation to stabilize the variances. See Section 10.7 for alternative approaches.

One approach to deciding upon a variance stabilizing transformation is to examine the plot of the absolute values of the residuals against the fitted values to find a function of $\hat{y}$ that serves as a rough upper-bound for the bulk of the plotted $|\hat{\varepsilon}|$. Two scenarios are presented here.

7.4.1 Power transformations

These are equivalent to those discussed for transformations to near normality. In fact, in [51, pp. 87–92] and [45, pp. 132–137], variance stabilizing and normalizing transformations are treated together, and the Box–Cox procedure is cited as one approach to determining appropriate variance stabilizing power transformations.

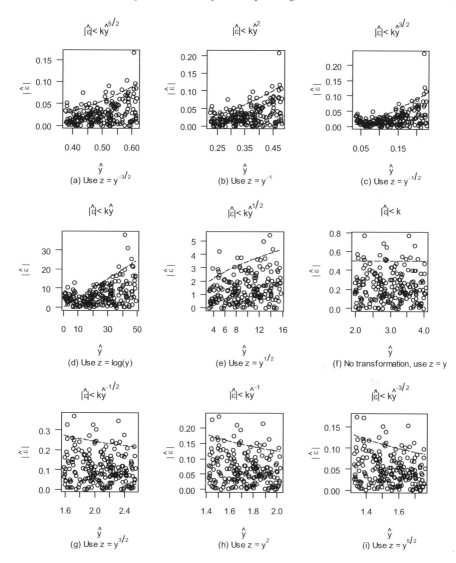

FIGURE 7.9: Diagnostic plots and examples of variance stabilizing transformations of the response variable.

Here, a visual approach is presented which appears in most texts on the subject; see, for example, [49, pp. 379–381], [50, pp. 174–175] and [45, p. 132] for brief discussions. The diagnostic plots in Figure 7.9 show some illustrations of choosing variance stabilizing transformations.

It should be remembered that finding an exact bounding function is usually not easy, nor a requirement for the process since exact fixes are not a necessity.

The trial-and-error approach mentioned for normalizing transformations works just as effectively here. In general:

- If the bounding curve is increasing and concave up, use $\lambda < 0$; see plots (a) through (c).

- If the bounding curve is increasing and linear, use $Z = \ln(Y)$; see plot (d).

- If the bounding curve is nondecreasing and concave down, use $0 < \lambda < 1$; see plot (e).

- If the bounding curve is horizontal, no transformation is needed; see plot (f).

- If the bounding curve is decreasing and concave up, use $\lambda > 1$; see plots (g) through (i).

Here is a brief discussion along the lines of [51, pp. 87–88] on how the transformations suggested in Figure 7.9 were arrived at.

Suppose a plot, as shown previously, suggests a bound of the form $|\hat{\varepsilon}| \leq k\,\hat{y}^a$, implying $\mathrm{var}(Y) \propto [\mathrm{E}(Y)]^{2a}$. Suppose further that a transformation $Z = Y^\lambda$ exists such that if $\mathrm{var}(Y) \propto [\mathrm{E}(Y)]^{2a}$, then $\mathrm{var}(Z) \propto$ constant. In [51], it is stated that it can be shown[5]

$$\mathrm{var}(Z) \propto [\mathrm{E}(Y)]^{2(\lambda+a-1)} .$$

In order to achieve the desired end result, $\mathrm{var}(Z) \propto$ constant, it is then a simple matter of solving $\lambda + a - 1 = 0$.

As an illustration, consider the case of Figure 7.9.b and suppose $|\hat{\varepsilon}| \leq k\,\hat{y}^2$ is a reasonably accurate bounding curve for the bulk of the plotted points. Then $\mathrm{var}(Y) \propto [\mathrm{E}(Y)]^{2(2)}$, suggesting the need for $\lambda + 1 = 0$ to be the case. So, the appropriate variance stabilizing transformation is $Z = Y^{-1}$.

7.4.2   The arcSine transformation

On occasion, the observed responses may represent binomial proportions with $0 \leq Y \leq 1$. For such cases, the $\hat{y}\,|\hat{\varepsilon}|$-plot will have the general appearance of Figure 7.10. The appropriate transformation for such cases is $Z = \arcsin\left(\sqrt{Y}\right)$.

Sample code for producing Figure 7.10 is given in the last section of this chapter.

---

[5] Details are not given and, curiously, this nifty idea is absent from the later edition of Montgomery's text with the same title. A more technical discussion on this matter is given in [58, pp. 364–368]; however, the level is beyond the scope of this Companion.

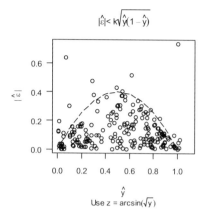

FIGURE 7.10: Diagnostic residual plot of data for which the arcsine transformation of the response might be appropriate.

7.5 Polynomial Regression

Scatterplots of the response variable Y against the explanatory variable X often suggest that Y may be expressed as a polynomial function of X. This gives rise to what is referred to as *polynomial regression*. Although such models really belong in the area of multiple linear regression, consider cooking up some bivariate data for which a model of the form $y_i = \beta_0 + \beta_1 x_i + \varepsilon_i$ is clearly inappropriate.

First, generate some x- and y-lists, and then store the data in a data frame. Call this data frame `PolyReg`.

```
x<-runif(50,0,5)
y<-round(-2*x^2+9*x+4+rnorm(50,0,1),1)
PolyReg<-data.frame("x"=x,"y"=y); rm(x,y)
```

Keep in mind that data generated will differ from run to run. The data `PolyReg` used for the following illustration was obtained after several runs and is stored in `Data07x04.R`.

A scatterplot of Y against X (see Figure 7.11.a) suggests a quadratic relationship exists between Y and X. Fitting the data to the model

$$y_i = \beta_0 + \beta_1 x_i + \beta_2 x_i^2 + \varepsilon_i$$

is accomplished using

```
quad.mod<-lm(y~x+I(x^2),PolyReg)
```

and the various residual plots, (b) though (d) in Figure 7.11, suggest no serious issues with the error terms assumptions. There may be three potential outliers in the residuals, but plot (a) does not suggest any points that seriously buck the general trend.

The model summary object `summary(quad.mod)` gives relevant information on the regression parameters and details needed to perform the goodness of fit test. To test

$$H_0 : y_i = \beta_0 + \varepsilon_i \qquad \{\text{Reduced model}\} \quad \text{vs.}$$
$$H_1 : y_i = \beta_0 + \beta_1 x_i + \beta_2 x_i^2 + \varepsilon_i \quad \{\text{Full model}\}$$

the last line of the summary output indicates the proposed model is significant (`p-value: < 2.2e-16`) with $R^2 = 0.9197$ and $R^2_{\text{Adj}} = 0.9163$.

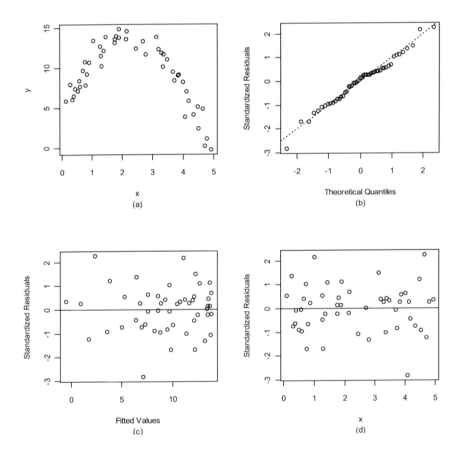

FIGURE 7.11: Diagnostic plots for the model obtained by fitting the dataset `PolyReg` to a quadratic model.

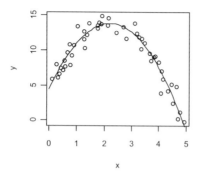

FIGURE 7.12: Plot of fitted quadratic model superimposed on scatterplot of dataset `PolyReg`.

The fitted model, with parameter estimates rounded to three decimal places, is $\hat{y} = 4.445 + 8.527x - 1.939x^2$ and a plot of the fitted model can be obtained (see Figure 7.12) using

```
xvals<-seq(0,5,.1)
yvals<-4.445+8.527*xvals-1.939*xvals^2
with(PolyReg,plot(y~x))
lines(xvals,yvals)
```

There are certain issues that are associated with polynomial regression models. See, for example, [45, p. 305] for further discussions. Estimation and prediction matters are left for later where multiple regression models are covered.

7.6 Piecewise Defined Models

On occasion, the fit of a model may be improved by partitioning the domain space (x-values) and defining the proposed model *piecewise*. For example, the mean response function for a model with one explanatory variable may have the form

$$F(X) = \begin{cases} f_1(X), & a_0 \leq X < a_1 \\ f_2(X), & a_1 \leq X \leq a_2 \end{cases}$$

where, in each f_k, $k = 1$ and 2, the regression parameters enter the model linearly.

Depending on the circumstances or the underlying principles, the functions f_k may be such that F is continuous or discontinuous. If the issue of continuity is not forced, the fitted model is not likely to be continuous.

Begin by generating some data to work with.

```
x1<-runif(40,0,1); x2<-runif(40,1,3)
y1<- -10*x1^2+10*x1+1+rnorm(40,mean=0,sd=.25)
y2<-x2+rnorm(40,mean=0,sd=.25)
x<-c(x1,x2); y<-c(y1,y2)
TwoPiece<-data.frame("x"=x,"y"=y);rm(x,y,x1,x2,y1,y2)
```

The data frame `TwoPiece` is stored in the file `Data07x05.R`.

As evident from Figure 7.13, the data should probably be fitted to a piecewise defined model. Two approaches are shown here. In both cases, an inspection of the xy-scatterplot is used to determine an appropriate partition of the domain space into subsets.

7.6.1   Subset regression

Since the left portion of Figure 7.13 suggests a quadratic model and the right a straight-line model, one might look at fitting the data to a model of the form

$$y_i = \begin{cases} \alpha_0 + \alpha_1 x_i + \alpha_2 x_i^2 + \varepsilon_i, & 0 \le x_i < 1 \\ \beta_0 + \beta_1 x_i + \varepsilon_i, & 1 \le x_i < 3 \end{cases}$$

Accomplishing this is a simple matter of partitioning the domain in the `lm` function call with the help of the `subset` argument.

```
piece1<-lm(y~x+I(x^2),TwoPiece,subset=(x<1))
piece2<-lm(y~x,TwoPiece,subset=(x>=1))
```

producing, to three decimal places,

$$\hat{y} = \begin{cases} 0.907 + 10.314\,x - 10.227\,x^2, & 0 \le x < 1 \\ 0.024 + 1.016\,x, & 1 \le x < 3 \end{cases}$$

Diagnostic procedures are then performed on each piece as before.

The graph of the fitted model (see Figure 7.13) superimposed on the scatterplot of the dataset is constructed as follows,

```
#Plot the data and a vertical line at x = 1
plot(y~x,TwoPiece,xlim=c(0,3),ylim=c(0,4.5))
abline(v=1,lty=3)
#Get coefficient estimates
a<-piece1$coefficients; b<-piece2$coefficients
```

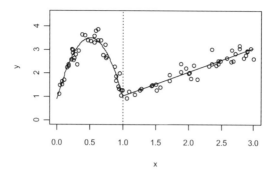

FIGURE 7.13: Plot of dataset `TwoPiece` with superimposed fitted model obtained using subset regression.

```
#Plot piece 1
xv<-seq(0,1,.01); yv<-a[1]+a[2]*xv+a[3]*xv^2; lines(xv,yv)
#Plot piece 2
segments(1,b[1]+b[2]*1,3,b[1]+b[2]*3)
```

Though not evident from the plot, the fitted model is discontinuous at $X = 1$. More often than not, this approach will produce discontinous fits.

7.6.2 Continuous piecewise regression

The discontinuity in the above fitted model can be addressed by using what is described as *broken stick regression* in [21, pp. 112–114]. The idea involved can be summarized briefly as follows: The two pieces are combined into a single function with the help of specially designed "switches" that turn on or off depending on what the inputted value of X is. If $X < 1$, the switch for piece 1 turns on and turns off for piece 2. Similarly, for $X \geq 1$ the switch for piece 1 turns off and turns on for piece 2. The design of the switches comes in to address the discontinuity, which is at $X = 1$. In order for the fitted model to approach the same value at $X = 1$ from the left *and* from the right, the use of what are called *basis functions* comes into play. For the current dataset, or any two-piece model with a discontinuity at $x = 1$, the basis functions take on the appearance

$$B_{\text{left}}(X) = \begin{cases} 1 - X, & X < 1 \\ 0, & X \geq 1 \end{cases} \quad \text{and} \quad B_{\text{right}}(X) = \begin{cases} 0, & X < 1 \\ X - 1, & X \geq 1 \end{cases}$$

The proposed model for the current data takes on the form

$$y_i = \beta_0 + \beta_1 B_{\text{left}}(x_i) + \beta_2 B_{\text{left}}^2(x_i) + \beta_3 B_{\text{right}}(x_i) + \varepsilon_i.$$

Observe that as X approaches 1 (from above or below) the model approaches the form $y_i = \beta_0 + \varepsilon_i$. Implementing the basis functions is made simple through the `ifelse` function.

```
bsl<-function(x) ifelse(x<1,1-x,0)
bsr<-function(x) ifelse(x<1,0,x-1)
```

Then the data are fitted to the proposed model using

```
bs.mod<-lm(y~bsl(x)+I(bsl(x)^2)+bsr(x),TwoPiece)
```

giving the fitted model

$$\hat{y} = 1.020 + 10.030\, B_{\text{left}}(x) - 10.135\, B_{\text{left}}^2(x) + 1.029\, B_{\text{right}}(x)$$

or

$$\hat{y} = \begin{cases} 0.915 + 10.24x - 10.135x^2, & 0 \le x < 1 \\ -0.009 + 1.029x, & 1 \le x < 3 \end{cases}$$

which is continuous at $x = 1$.

The fitted model can be viewed graphically (see Figure 7.14) using

```
plot(y~x,TwoPiece,xlim=c(0,3),ylim=c(0,5))
abline(v=1,lty=3)
b<-bs.mod$coefficients
xv<-seq(0,3,.01)
yv<-b[1]+b[2]*bsl(xv)+b[3]*bsl(xv)^2+b[4]*bsr(xv)
lines(xv,yv)
```

While Figure 7.14 does not look much different from Figure 7.13, a comparison of the estimated regression coefficients for the two approaches shows that there is a difference between the two fitted models.

7.6.3   A three-piece example

The two approaches described above can also be applied to data that suggest a partition of the domain space into more than two subsets. For example, suppose

$$F(x) = \begin{cases} f_1(x), & c_0 \le x < c_1 \\ f_2(x), & c_1 \le x < c_2 \\ f_3(x), & c_2 \le x \le c_3 \end{cases}$$

where, in each f_k, $k = 1, 2, 3$, the regression parameters enter the model linearly. For subset regression, the process is as described earlier and this works well as long as the number of observations in each subset are sufficiently large.

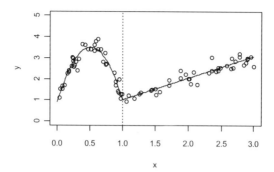

FIGURE 7.14: Plot of a continuous piecewise model obtained using broken-stick regression.

For continuous piecewise regression, the trick (and difficulty) is finding basis functions that force continuity in the fitted model. This is not necessarily as easy as the two-piece case described earlier. Consider a model having three linear piecewise defined functions and, hence, three linear basis functions. The mean response function can be represented by

$$F(x) = \beta_0 + \beta_1 B_1(x) + \beta_2 B_2(x) + \beta_3 B_3(x),$$

and the goal is to achieve continuity at all *knot points* (points that partition the domain space); that is, if the two partition points are $x = c_1$ and $x = c_2$, then the following is needed:

$$\lim_{x \to c_1^-} F(x) = \lim_{x \to c_1^+} F(x) \text{ and } \lim_{x \to c_2^-} F(x) = \lim_{x \to c_2^+} F(x).$$

Using

$$B_1(x) = \begin{cases} c_1 - x & \text{if } x < c_1, \\ 0 & \text{otherwise}; \end{cases}$$

$$B_2(x) = \begin{cases} 0 & \text{if } x < c_1, \\ x - c_1 & \text{if } c_1 \leq x < c_2, \quad \text{and} \\ c_2 - c_1 & \text{otherwise}; \end{cases}$$

$$B_3(x) = \begin{cases} c_2 - x & \text{if } x \geq c_2, \\ 0 & \text{otherwise}; \end{cases}$$

it is seen that

$$\lim_{x \to c_1^-} f(x) = \beta_0 = \lim_{x \to c_1^+} f(x), \quad \text{and}$$

$$\lim_{x \to c_2^-} f(x) = \beta_0 + \beta_2 (c_2 - c_1) = \lim_{x \to c_2^+} f(x).$$

To illustrate this, first generate some data:

```
x1<-runif(30,0,1); x2<-runif(30,1,2); x3<-runif(30,2,3)
y1<--5+2*x1+rnorm(30,mean=0, sd=.1)
y2<--7-(x2-1)+rnorm(30,mean=0, sd=.1)
y3<--6+1/2*(x3-2)+rnorm(30,mean=0, sd=.1)
x<-c(x1,x2,x3); y<-c(y1,y2,y3)
ThreePiece<-data.frame(cbind(x,y))
rm(x1,x2,x3,y1,y2,y3,x,y)
```

The dataset `ThreePiece` used for the following is stored in the file `Data07x06.R`.

By construction, the *knots* are at $x = 1$ and $x = 2$, so $c_1 = 1$ and $c_2 = 2$. The basis functions are defined using the `ifelse` function and the data are then fitted to the model

$$y_i = \beta_0 + \beta_1 B_1(x_i) + \beta_2 B_2(x_i) + \beta_3 B_3(x_i) + \varepsilon_i.$$

using

```
bs1<-function(x) {ifelse(x<1,1-x,0)}
bs2<-function(x) {ifelse(x<1,0,ifelse(x<2,x-1,1))}
bs3<-function(x) {ifelse(x<2,0,2-x)}
bs3.mod<-lm(y~bs1(x)+bs2(x)+bs3(x),ThreePiece)
```

Figure 7.15 was obtained using the code shown below:

```
b<-bs3.mod$coefficients
with(ThreePiece,plot(x,y))
xv<-seq(0,3,.01)
yv<-b[1]+b[2]*bs1(xv)+b[3]*bs2(xv)+b[4]*bs3(xv)
lines(xv,yv)
```

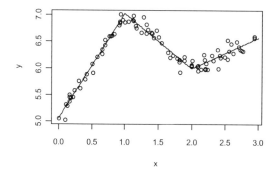

FIGURE 7.15: Plot of dataset `ThreePiece` along with fitted piecewise model obtained using broken-stick regression.

On occassion, the data cannot be partitioned by values of the explanatory variable. However, a natural grouping (not necessarily by magnitude) may exist in the observed responses. This leads to another possible strategy.

7.7   Introducing Categorical Variables

In some situations, usually where replicates of the response variable Y at each level of the explanatory variables are involved, it might be possible, or appropriate, to separate the observed responses into two or more categories.

Very often the same values for X are used for all q factor levels, and usually a single observed response, y_{ij}, is obtained at the ith level of X and the jth level of the grouping factor. This may or may not always be the case, but it is typically desirable. This scenario is computationally analogous to *analysis of covariance* models discussed in Chapter 12.

Data for such models might be tabulated in the following form:

	Factors			
X	1	2	$\cdots$	q
x_1	y_{11}	y_{12}	$\cdots$	y_{1q}
x_2	y_{21}	y_{22}	$\cdots$	y_{2q}
$\vdots$	$\vdots$	$\vdots$	$\ddots$	$\vdots$
x_n	y_{n1}	y_{n2}	$\cdots$	y_{nq}

Two simple scenarios are presented here.

7.7.1   Parallel straight-line models

Consider having a single continuous explanatory variable X and a factor with q levels (a categorical variable with q categories). Assume a linear relationship between the response variable Y and the explanatory variable X, and assume equal slopes across factor levels. This model can be written in the form

$$
y_{ij} = \begin{cases}
\beta_0 + \alpha_1 + \beta_1 x_i + \varepsilon_{i1} & \text{for } j = 1 \text{ and } i = 1, 2, \ldots, n \\
\beta_0 + \alpha_2 + \beta_1 x_i + \varepsilon_{i2} & \text{for } j = 2 \text{ and } i = 1, 2, \ldots, n \\
\quad \vdots & \quad \vdots \\
\beta_0 + \alpha_q + \beta_1 x_i + \varepsilon_{iq} & \text{for } j = q \text{ and } i = 1, 2, \ldots, n
\end{cases}
$$

or more concisely as

$$
y_{ij} = \beta_0 + \alpha_j + \beta_1 x_i + \varepsilon_{ij}
$$

for $i = 1, 2, \ldots, n$ and $j = 1, 2, \ldots, q$. In a very real sense, the model represents a collection of separate simple regression models having the same domain space and tied together in a single model by means of the parameters α_j.

For purposes of illustration, first generate some data:

```
x<-runif(30,0,2); y1<-1+5*x+rnorm(30,sd=.35)
y2<-3+5*x+rnorm(30,sd=.35); y3<-5+5*x+rnorm(30,sd=.35)
x<-c(x,x,x); y<-c(y1,y2,y3)
Category<-c(rep(1,30),rep(2,30),rep(3,30))
CatReg1<-data.frame(x,y,Category)
rm(x,y1,y2,y3,y,Category)
CatReg1$Category<-factor(CatReg1$Category)
```

The dataset `CatReg1`, used in the following illustration, is stored in the file `Data07x07.R`.

The inclusion of an additional argument in the following `plot` command enables the plotting of different symbols for each category as shown in Figure 7.16.

```
with(CatReg1,plot(x,y,pch=unclass(Category)))
```

The `pch` argument, with the help of the function `unclass`, instructs R as to the type of symbol to be used to plot points. The fact that the category levels are numbers helps since plotting symbols are identified by "1", "2", etc. The three distinct linear trends in the plot indicate the need for including a categorical explanatory variable with three levels. A scatterplot of Y against X without using different symbols for each category might show nothing unusual except for an appearance of a large error variance, or it might be that a noticeable

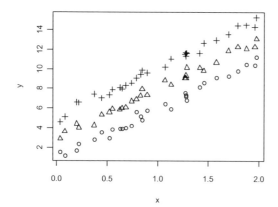

FIGURE 7.16: Scatterplot of dataset `CatReg1` showing distinct parallel linear trends. Plotted points are identified by the use of distinct symbols.

separation of the plotted points into trended subsets exists as is evident in Figure 7.16.

In fitting the data to a model one might, just to be on the safe side, wish to see if the slopes of the three trends are significantly different. This would be characterized by interaction between the numerical and categorical variable, and can be tested as follows:

The task is to test

$$H_0 : y_{ij} = \beta_0 + \alpha_j + \beta_1 x_i + \varepsilon_{ij} \qquad \text{\{Reduced model\} vs.}$$
$$H_1 : y_{ij} = \beta_0 + \alpha_j + \beta_1 x_i + (\alpha\beta)_{j1} \, x_i + \varepsilon_{ij} \text{ \{Full model\}}$$

These hypotheses can be tested using using the commands

```
> parallel.mod<-lm(y~x+Category,CatReg1)
> nonpar.mod<-lm(y~x+Category+x:Category,CatReg1)
> anova(parallel.mod,nonpar.mod)
Analysis of Variance Table
Model 1: y ~x + Category
Model 2: y ~x + Category + x:Category
  Res.Df    RSS  Df  Sum of Sq       F  Pr(>F)
1     86  10.020
2     84  10.006   2   0.013811  0.058  0.9437
```

There are two points to note here. First, x:Category represents the interaction component in the non-parallel model, and the fitting command for this model can be condensed to

```
nonpar.mod<-lm(y~x*Category,CatReg1)
```

Second, the function anova() compares the two models using a *partial F-test*. More uses of partial F-tests appear in later chapters.

The last line of the output suggests that there is insufficient evidence to support the full model; hence, the constant slope model is assumed and parallel.mod contains

```
Coefficients:

(Intercept)          x   Category2   Category3
     0.8934   5.0475      2.0370      4.1388
```

which gives the parameter estimates for the parallel model. To interpret this output, recall that the constant slope model mentioned earlier has the form

$$y_{ij} = \begin{cases} \beta_0 + \alpha_1 + \beta_1 x_i + \varepsilon_{i1} \text{ if } j = 1 \\ \beta_0 + \alpha_2 + \beta_1 x_i + \varepsilon_{i2} \text{ if } j = 2 \\ \beta_0 + \alpha_3 + \beta_1 x_i + \varepsilon_{i3} \text{ if } j = 3 \end{cases}$$

Thus, parameter estimates for the fitted model are

$$\hat{\beta}_0 = 0.8934, \quad \hat{\beta}_1 = 5.0475, \quad \hat{\alpha}_1 = 0, \quad \hat{\alpha}_2 = 2.0370, \quad \text{and} \quad \hat{\alpha}_3 = 4.1388,$$

where $\alpha_1 = 0$ is the default value assigned by R in the fitting process. More on why this is done appears in Chapter 11. Consequently, the fitted regression model is

$$\hat{y} = \begin{cases} 0.8934 + 5.0475x & \text{for level 1} \\ 2.9304 + 5.0475x & \text{for level 2} \\ 5.0322 + 5.0475x & \text{for level 3} \end{cases}$$

Lines representing the fitted model for the three factor levels along with a legend can be superimposed on the scatterplot (see Figure 7.17) by attaching the following code to the end of the earlier plotting command for Figure 7.16:

```
#Extract parameter estimates
bs<-parallel.mod$coefficients
#Line for level 1
abline(a=bs[1],b=bs[2],lty=1)
#Line for level 2
abline(a=bs[1]+bs[3],b=bs[2],lty=2)
```

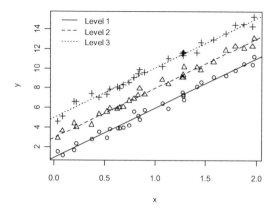

FIGURE 7.17: Scatterplot of dataset `CatReg1` with plotted points identified by category along with fitted regression lines for each category.

```
#Line for level 3
abline(a=bs[1]+bs[4],bs[2],lty=3)
#Include legend
legend(0,16,bty="n",
    legend=c("Level 1","Level 2","Level 3"),lty=c(1,2,3))
```

Clear the workspace before continuing.

7.7.2 Non-parallel straight-line models

If the null hypothesis in the previous example had been rejected, then interaction between the categorical and explanatory variable would have been considered significant. Graphically, this is indicated by a difference between slopes (rates of change) across levels of the categorical variable. A model of this form can be written as

$$y_{ij} = \begin{cases} \beta_0 + \alpha_1 + [\beta_1 + (\alpha\beta)_{11}] x_i + \varepsilon_{i1} & \text{for } j = 1 \text{ and } i = 1, 2, \ldots, n \\ \beta_0 + \alpha_2 + [\beta_1 + (\alpha\beta)_{21}] x_i + \varepsilon_{i2} & \text{for } j = 2 \text{ and } i = 1, 2, \ldots, n \\ \quad \vdots & \quad \vdots \\ \beta_0 + \alpha_q + \left[\beta_1 + (\alpha\beta)_{q1}\right] x_i + \varepsilon_{iq} & \text{for } j = q \text{ and } i = 1, 2, \ldots, n \end{cases}$$

or more concisely as

$$y_{ij} = \beta_0 + \alpha_j + \beta_1 x_i + (\alpha\beta)_{j1} x_i + \varepsilon_{ij}$$

for $i = 1, 2, \ldots, n$ and $j = 1, 2, \ldots, q$. As for the previous case, the model represents a collection of separate simple regression models having the same domain space and tied together in a single model by the parameters α_j and $(\alpha\beta)_{j1}$.

Again, consider a simple case in which the response is linearly related to a single numerical explanatory variable but provides for slopes that vary across factor levels. For the following illustration, tweak the previous data generating commands so that the response functions across factor levels have different slopes.

```
x<-runif(30,0,2); y1<-1+5*x+rnorm(30,sd=.25)

y2<-3+3*x+rnorm(30,sd=.25); y3<-5+2*x+rnorm(30,sd=.25)

x<-c(x,x,x); y<-c(y1,y2,y3)

Category<-c(rep(1,30),rep(2,30),rep(3,30))

CatReg2<-data.frame(x,y,Category)

rm(x,y1,y2,y3,y,Category)

CatReg2$Category<-factor(CatReg2$Category)
```

The dataset `CatReg2` used in the following illustration is stored in the file `Data07x08.R`.

Going straight to testing the hypotheses

$$\text{H}_0 : y_{ij} = \beta_0 + \alpha_j + \beta_1 x_i + \varepsilon_{ij} \qquad \{\text{Reduced model}\} \text{ vs.}$$
$$\text{H}_1 : y_{ij} = \beta_0 + \alpha_j + \beta_1 x_i + (\alpha\beta)_{j1} x_i + \varepsilon_{ij} \{\text{Full model}\}$$

for nonconstant slopes,

```
> parallel.mod<-lm(y~x+Category,CatReg2)

> nonpar.mod<-lm(y~x+Category+x:Category,CatReg2)

> anova(parallel.mod,nonpar.mod)

Analysis of Variance Table

Model 1: y ~x + Category

Model 2: y ~x + Category + x:Category
   Res.Df     RSS  Df  Sum of Sq       F      Pr(>F)
1      86  62.124
2      84   5.836   2     56.288  405.08   < 2.2e-16 ***
```

The last line of the output indicates there is sufficient evidence to support the full model. Parameter estimates for `nonpar.mod` are found from

```
> round(nonpar.mod$coefficients,3)
(Intercept)              x    Category2   Category3
      1.032          4.963        1.990       4.073
x:Category2   x:Category3
     -1.914        -3.065
```

So,

$$\hat{\beta}_0 = 1.032, \quad \hat{\beta}_1 = 4.963, \quad \hat{\alpha}_2 = 1.990, \quad \hat{\alpha}_3 = 4.073,$$
$$(\alpha\beta)_{21} = -1.914, \quad (\alpha\beta)_{31} = -3.065$$

and $\alpha_1 = (\alpha\beta)_{11} = 0$ are default values assigned by R. The fitted regression model is

$$\hat{y} = \begin{cases} 1.032 + 4.963x & \text{for level 1} \\ 3.022 + 3.049x & \text{for level 2} \\ 5.105 + 1.898x & \text{for level 3} \end{cases}$$

and the plot of the resulting fitted model in Figure 7.18 can be obtained using code of the form

```
with(CatReg2,plot(x,y,pch=unclass(Category)))
bs<-nonpar.mod$coefficients
abline(a=bs[1],b=bs[2],lty=1)
abline(a=bs[1]+bs[3],b=bs[2]+bs[5],lty=2)
abline(a=bs[1]+bs[4],b=bs[2]+bs[6],lty=3)
legend(0,11,bty="n",
    legend=c("Level 1","Level 2","Level 3"),lty=c(1,2,3))
```

Categorical variables can also be used to construct piecewise defined models in which the categories are defined by partitions of the domain space. The resulting fitted models very closely resemble those obtained via subset regression. However, in using a categorical variable, the fitted model is treated as a single model rather than three (or more) separate models, as is the case with subset regression. This distinction of a "single model" versus "separate models" may or may not be useful to consider during the model construction phase.

7.8 For the Curious

The Box–Cox procedure can be viewed as an interesting programming exercise. Sample code for this exercise along with a demonstration of two other features of R — inserting mathematical annotations in figures and the split function — are illustrated here.

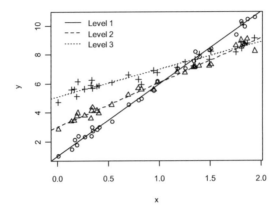

FIGURE 7.18: Scatterplot of dataset `CatReg2` illustrating a clear separation of plotted points by three non-parallel trends, along with superimposed fitted lines.

7.8.1 The Box–Cox procedure revisited

First clear the workspace, then reload the dataset `RightSkewed` using a command of the form

```
source("z:/Docs/RCompanion/Chapter7/Data/Data07x03.R")
```

It was mentioned in Section 7.2 that the procedure uses a modified power transformation

$$z = \begin{cases} (y^\lambda - 1)/\lambda & \text{if } \lambda \neq 0 \\ \ln(y) & \text{if } \lambda = 0, \end{cases}$$

and if $\lambda = 1$, a power transformation is not performed. To determine the most appropriate choice of λ, the goal is to maximize the likelihood function

$$l(\lambda) = -\frac{n}{2} \ln\left[\frac{1}{n}\sum_{i=1}^{n}(z_i - \hat{z}_i)^2\right] + (\lambda - 1)\sum_{i=1}^{n}\ln(y_i),$$

where $\hat{z}_i$ is the ith fitted value for the model

$$z_i = \beta_0 + \beta_1 x_i + \varepsilon_i,$$

See, for example, [42, p. 207] or [45, pp. 134–137].

Code to facilitate a graphical approximation of λ (see Figure 7.19) that maximizes $l(\lambda)$ might be as follows:

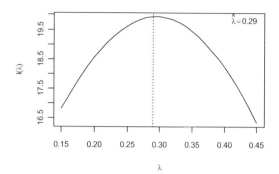

FIGURE 7.19: Plot of $l = l(\lambda)$ using the Box–Cox procedure to obtain an estimate $\hat{\lambda}$ to use in a power transformation of the response in dataset RightSkewed.

```
#For convenience
attach(RightSkewed)
#Define transformation function
z<-function(Y,a){
    if(abs(a)<=0.001){log(Y)} else {(Y^a-1)/a}}
#Obtain sample size and sequence of lambdas
n<-length(y); l<-seq(.15,.45,0.01)
#Initialize likelihood function vector
L<-numeric(length(l))
#Compute likelihood values
for (k in 1:length(l)){
    e<-residuals(lm(z(y,l[k])~x))
    L[k]<- -n/2*log(sum(e^2)/n)+(l[k]-1)*sum(log(y))}
#Set up window for plotting
win.graph(width=4,height=3)
par(mfrow=c(1,1),ps=10,cex=.75)
#Obtain plot of likelihood function against lambda
plot(L~l,type="l",xlab=expression(lambda),
        ylab=expression(l(lambda)))
#Mark maximizing lambda
abline(v=l[L==max(L)],lty=3)
```

```
#Insert maximizing lambda value on figure
text(max(l),max(L),
    bquote(hat(lambda)==.(round(l[L==max(L)],2))),adj=1)
detach(RightSkewed)
```

The plot of $l = l(\lambda)$ in Figure 7.19 is a scaled vertical translation of Figure 7.7, obtained using the function boxcox from package MASS. However, observe the maximizing values of $\hat{\lambda}$ for the two match up.

7.8.2 Inserting mathematical annotation in figures

An illustration of inserting mathematical annotation in Figures 7.9 and 7.10 appears in the following code used to generate Figure 7.10:

```
#The first two lines generate the data
x<-runif(200,0,pi/2)
y<-sin(x+rnorm(200,mean=0,sd=0.25))^2
#Fit the data to a model
mod<-lm(y~x)
#Plot the points, set tick mark label orientation
with(mod,plot(fitted.values,abs(residuals),las=1,
#Insert the main title with mathematical annotation
 main=expression(paste(abs(hat(epsilon))<
                k*sqrt(hat(y)*(1-hat(y))))),
#Set x-limits and insert x-label
   xlim=c(0,1.1),xlab=expression(hat(y)),
#Insert y-label and subtitle with mathematical annotation
   ylab=expression(paste(abs(hat(epsilon)))),
   sub=expression(paste("Use ",z==arcsin(sqrt(y))))))
#Plot the bounding curve (eye-balling the scale)
xv<-seq(0,1,.01); yv<-1.6*xv*(1-xv); lines(xv,yv,lty=5)
```

Much of the syntax for inserting mathematical annotation is the same as for LaTeX, which is a plus for those familiar with LaTeX. See the R documentation page for plotmath in package grDevices for a detailed list of what can be inserted in a graphics image, and how.

7.8.3   The split function

The split function is somewhat similar to the earlier seen cut function in that it permits a grouping of the data. However, while the cut function groups the data by the domain space, the split function makes use of categorical groupings of the response. The plot of the dataset CatReg1 in Figure 7.16 can also be obtained using the code

```
with(CatReg1,plot(x,y,type="n"))
xs<-with(CatReg1,split(x,Category))
ys<-with(CatReg1,split(y,Category))
for (i in 1:3){points(xs[[i]],ys[[i]],pch=i)}
```

Like the cut function, this function might find uses where it may be desirable to look at only portions of the data by some grouping factor.

Chapter 8

Multiple Linear Regression

8.1   Introduction 167
8.2   Exploratory Data Analysis 169
8.3   Model Construction and Fit 171
8.4   Diagnostics 173
     8.4.1   The constant variance assumption 174
         8.4.1.1   F-test for two population variances 176
         8.4.1.2   The Brown–Forsyth test 177
     8.4.2   The normality assumption 177
         8.4.2.1   QQ normal probability correlation coefficient test 178
         8.4.2.2   Shapiro–Wilk test 178
     8.4.3   The independence assumption 178
     8.4.4   The presence and influence of outliers 179
         8.4.4.1   Outlying observed responses 179
         8.4.4.2   Outliers in the domain space 181
         8.4.4.3   Influence of outliers on corresponding fitted values 181
         8.4.4.4   Influence of outliers on all fitted values 182
         8.4.4.5   Influence of outliers on parameter estimates 182
         8.4.4.6   Quick influence analysis 184
8.5   Estimating Regression Parameters 186
     8.5.1   One-at-a-time t-intervals 186
     8.5.2   Simultaneous t-intervals 186
     8.5.3   Simultaneous F-intervals 187
8.6   Confidence Intervals for the Mean Response 188
     8.6.1   One-at-a-time t-intervals 188
     8.6.2   Simultaneous t-intervals 189
     8.6.3   Simultaneous F-intervals 190
8.7   Prediction Intervals for New Observations 192
     8.7.1   t-interval for a single new response 192
     8.7.2   Simultaneous t-intervals 192
     8.7.3   Simultaneous F-intervals 194
8.8   For the Curious 195
     8.8.1   Fitting and testing a model from scratch 195
     8.8.2   Joint confidence regions for regression parameters 197
     8.8.3   Confidence regions for the mean response 199
     8.8.4   Prediction regions 202

8.1   Introduction

Let y_i, $i = 1, 2, \ldots, n$ represent *observed values* for a continuous *response variable* Y. For $j = 1, 2, \ldots, p$, let x_{ij}, denote the corresponding values for the continuous *explanatory variables* $X_1, X_2, \ldots, X_p$ and denote the unknown *regression parameters* by β_j. Finally, let ε_i represent the *random error terms* corresponding to each observed response. As with simple linear regression, the explanatory variables are assumed to be measured without error.

The general structure of a *multiple linear regression model* in *algebraic form* has the appearance

$$y_i = \beta_0 + \beta_1 x_{i1} + \cdots + \beta_p x_{ip} + \varepsilon_i.$$

As with the simple linear regression model, it is assumed that for each $i = 1, 2, \ldots, n$ the error terms, ε_i, have constant variance σ^2, are independent, and are identically and normally distributed with $\varepsilon_i \sim N(0, \sigma^2)$.

In matrix form, the model also has the appearance

$$\mathbf{y} = \mathbf{X}\boldsymbol{\beta} + \boldsymbol{\varepsilon}.$$

Again, it is assumed that the *design matrix*, $\mathbf{X}$, has *full column rank*; the *response vector* $\mathbf{y}$ is a solution of $\mathbf{y} = \mathbf{X}\boldsymbol{\beta} + \boldsymbol{\varepsilon}$; and the entries of the *error vector*, $\boldsymbol{\varepsilon}$, satisfy the above-mentioned conditions. Note also that the response variable is assumed to be related (in the statistical sense) to a combination of at least two continuous explanatory variables in a manner such that the corresponding regression parameters enter the model linearly.

The data for all illustrations in this chapter (see Table 8.1) are stored in the data frame `MultipleReg` and saved in the file `Data08x01.R`.

TABLE 8.1: Data for Chapter 8 Illustrations

Y	X_1	X_2	X_3	Y	X_1	X_2	X_3
15.09	9.83	1.87	8.14	29.38	3.14	2.82	11.32
34.37	6.62	3.94	14.16	3.30	8.09	1.72	2.04
24.64	5.27	4.95	9.17	16.17	6.76	4.57	6.53
19.16	1.96	4.78	5.38	29.24	7.59	2.43	12.55
26.68	6.47	2.87	11.20	48.00	4.70	4.54	17.83
38.04	9.02	1.11	16.99	39.97	8.18	4.15	16.70
5.59	9.32	1.09	4.01	17.16	1.00	2.72	4.99
29.31	9.80	1.06	13.55	13.80	9.24	3.10	6.59
28.27	7.89	4.59	11.09	22.86	8.05	3.87	9.81
7.04	2.45	3.18	1.31	30.05	3.87	4.73	11.07
38.56	5.73	2.29	15.72	16.50	7.83	1.36	7.70
3.95	8.91	2.01	3.73	40.04	7.48	4.53	16.08
1.57	7.95	2.16	1.98	2.90	6.06	2.73	1.61
10.38	6.09	4.60	3.34	42.00	5.81	1.72	16.86
47.61	3.28	2.80	17.63	39.08	9.69	4.26	16.43
12.71	4.28	3.23	4.87	24.74	7.01	1.81	10.18
28.19	1.56	2.42	10.12	34.61	1.53	3.74	12.50
6.80	5.24	2.46	1.79	45.54	7.80	2.14	19.27
26.25	3.48	2.48	9.82	5.60	6.90	4.89	1.52
45.33	6.91	4.73	17.68	23.70	3.51	4.67	7.53

The data frame `MultipleReg` was generated with the help of the functions `runif` and `rnorm` in much the same way as earlier datasets were generated.

8.2 Exploratory Data Analysis

An initial graphical analysis might focus on the distributional properties of Y; the relationships between Y and each of X_1, X_2, and X_3; the relationships between pairs of X_1, X_2, and X_3; and the presence of unusual observations. One might begin with a histogram of the observed responses, the earlier written function `snazzy.hist` can be used.[1]

```
with(MultipleReg,snazzy.hist(y))
```

(see Figure 8.1). Histograms *might* provide some information on the symmetry and spread of the observed responses in relation to the normal distribution having the same mean and standard deviation as those of the sample data.

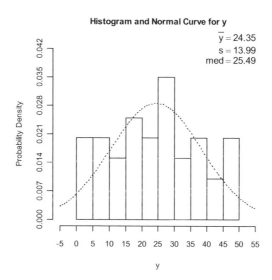

FIGURE 8.1: Density histogram and normal curve for observed responses in `MultipleReg` dataset.

---

[1] This would have to be sourced; use code of the form

```
source("z:/Docs/RCompanion/Chapter5/Functions/snazzyhist.R")
```

Boxplots can also be used as they provide information on symmetry and the presence of outliers (*in the univariate sense*); however, a caution is appropriate here. When combining boxplots on a single figure, pay attention to scale (and units). It is preferable to use separate plots if there is a large difference in ranges between variables, and it is definitely preferable to use separate plots if variable units differ.

A *matrix scatterplot* can be used to look for patterns or surprising behavior in both the response and the explanatory variables (see Figure 8.2). One of the ways to obtain this figure is

```
with(MultipleReg,pairs(cbind(y,x1,x2,x3)))
```

Another way is to simply enter `plot(MultipleReg)`.

An examination of (X_j, Y) pairs in scatterplots such as Figure 8.2 provide some (often vague) information on issues associated with approximate relat-

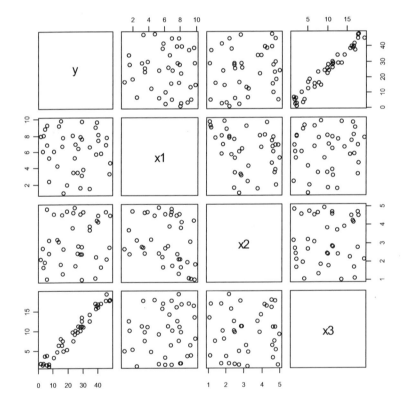

FIGURE 8.2: Matrix scatterplot of `MultipleReg` dataset illustrating pairwise relationships for all variables in the dataset.

ional fits between the response and explanatory variables. The more useful information obtainable from such plots concerns the presence of possible linear relationships between pairs of explanatory variables, (X_j, X_k), where $j \neq k$.

The `summary` function can be used here if summary statistics for the data are desired, the execution being as for the simple regression case.

8.3 Model Construction and Fit

The function `lm` is used to fit the data to the proposed model,

```
Multiple.mod<-lm(y~x1+x2+x3,MultipleReg)
```

and the contents of `Multiple.mod` include information analogous to the simple regression case.

```
> round(Multiple.mod$coefficients,4)
(Intercept)        x1        x2        x3
     2.9411   -0.9098    0.8543    2.4917
```

So, the fitted model is $\hat{y} = 2.9411 - 0.9098x_1 + 0.8543x_2 + 2.4917x_3$.

Recall that in Chapter 6 it was stated that the `anova` function, as used for simple regression, will not provide the traditional ANOVA table for a multiple regression model. To see that this is really the case execute, for example, the two commands

```
anova(lm(y~x1+x2+x3,MultipleReg))
anova(lm(y~x2+x3+x1,MultipleReg))
```

The `F value` for any one variable is dependent on the order in which it appears in the formula used to fit the data to the model. Here is a simplified explanation of what the row corresponding to x2 represents in the two `anova` calls above.

For the first call, the formula `y~x1+x2+x3` instructs the `anova` function to look at the hypotheses

$$H_0 : y_i = \beta_0 + \beta_1 x_{i1} + \varepsilon_i \qquad \text{\{Reduced model\} vs.}$$
$$H_1 : y_i = \beta_0 + \beta_1 x_{i1} + \beta_2 x_{i2} + \varepsilon_i \qquad \text{\{Full model\}}$$

So, the row corresponding to x2 addresses the contribution significance of x2 to the model containing x1 and x2, given that x1 is in the reduced model.

For the second call, the formula `y~x2+x3+x1` instructs the `anova` function to look at the hypotheses

$$H_0 : y_i = \beta_0 + \varepsilon_i \qquad \{\text{Reduced model}\} \text{ vs.}$$
$$H_1 : y_i = \beta_0 + \beta_2 x_{i2} + \varepsilon_i \qquad \{\text{Full model}\}$$

and in this case the row corresponding to x2 looks at the contribution significance of x2 to the model containing *only* x2, given that there are no variables in the reduced model.

The contribution significance of each variable, and the overall model fit is best obtained from the model summary object.

```
Multiple.sum<-summary(Multiple.mod)
```

Take a look at the contents of the object `Multiple.sum` with the help of the function `names`. Then,

```
> Multiple.sum$coefficients
```

	Estimate	Std. Error	t value	Pr(>|t|)
(Intercept)	2.9410671	0.66798813	4.402873	9.159731e-05
x1	-0.9098496	0.06287012	-14.471892	1.405530e-16
x2	0.8543410	0.12738684	6.706666	7.987807e-08
x3	2.4917171	0.02744574	90.787012	4.094410e-44

provides information on the *contribution significance* of each variable to the *proposed (full) model*, given that the other two variables are in the reduced model. For example, the row corresponding to x1 looks at the hypotheses

$$H_0 : y_i = \beta_0 + \beta_2 x_{i2} + \beta_3 x_{i3} + \varepsilon_i \qquad \{\text{Reduced model}\} \text{ vs.}$$
$$H_1 : y_i = \beta_0 + \beta_1 x_{i1} + \beta_2 x_{i2} + \beta_3 x_{i3} + \varepsilon_i \quad \{\text{Full model}\}.$$

If simply `Multiple.sum` is entered, the output includes the lines

```
Residual standard error: 0.9442 on 36 degrees of freedom
Multiple R-squared: 0.9958, Adjusted R-squared: 0.9954
F-statistic: 2843 on 3 and 36 DF, p-value: < 2.2e-16
```

which provide quite a bit of additional information.

Begin with the third line of the output shown above. This provides results of the *goodness of fit F-test*, or the model significance test involving the hypotheses

$$H_0 : y_i = \beta_0 + \varepsilon_i \qquad \{\text{Reduced model}\} \text{ vs.}$$
$$H_1 : y_i = \beta_0 + \beta_1 x_{i1} + \beta_2 x_{i2} + \beta_3 x_{i3} + \varepsilon_i \quad \{\text{Full model}\}$$

The second line of the `Multiple.sum` output shown previously contains the *coefficient of multiple determination* (R^2), `Multiple R-squared`, and the *adjusted coefficient of multiple determination* (R^2_{Adj}), `Adjusted R-squared`. Finally, the first line gives the *residual standard error* (s) along with its degrees of freedom.

Because of the presence of multiple explanatory variables, there may be a need to select the "best" variables, and there may be a need to select the "best" model from a pool of candidate models. Illustrations of these procedures are addressed in Chapter 9. For the present illustration, assume the model

$$y_i = \beta_0 + \beta_1 x_{i1} + \beta_2 x_{i2} + \beta_3 x_{i3} + \varepsilon_i$$

is the only model to be worked with.

8.4 Diagnostics

There is quite a bit more involved in the diagnostics stage for multiple regression models and this can require some preparatory work. The objects of interest that are contained in the fitted model object `Multiple.mod` include coefficients ($\hat{\boldsymbol{\beta}}$), residuals ($\hat{\boldsymbol{\varepsilon}}$), and `fitted.values` ($\hat{\mathbf{y}}$). Of use in the model summary object `Multiple.sum` are sigma (s), the degrees of freedom of s^2 which is contained in `df[2]`, and `cov.unscaled`, which is the matrix $(\mathbf{X'X})^{-1}$. These objects will be used to obtain the various statistics needed for the diagnostics to be performed.

In addition to the usual (*unstandardized*) *residuals*, $\hat{\varepsilon}_i$, available in `Multiple.mod`, two transformed versions of the residuals may find use in the preliminary stages of the following diagnostics: *standardized* (or *semi-studentized*) *residuals*,

$$\hat{r}_i = \frac{\hat{\varepsilon}_i}{s},$$

and *studentized* (or *internally studentized*) *residuals*,

$$\hat{r}_i^* = \frac{\hat{\varepsilon}_i}{s\sqrt{1 - h_{ii}}}.$$

The h_{ii} terms in the above formula are called *leverages* and are the diagonal entries of the *hat-matrix* $\mathbf{H} = \mathbf{X}(\mathbf{X'X})^{-1}\mathbf{X'}$. These find use in the analysis of explanatory variable *p-tuples*, $(x_{i1}, x_{i2}, \ldots, x_{ip})$, for flagging potential outlying cases.

One further form of residuals, seen earlier, that plays a role in outlier analysis is the *studentized deleted* (or *externally studentized*) *residuals*,

$$\hat{d}_i^* = \hat{\varepsilon}_i \sqrt{\frac{n - p - 2}{s^2(n - p - 1)(1 - h_{ii}) - \hat{\varepsilon}_i^2}}.$$

The choice of which form of residuals to use in the diagnostics process is sometimes guided by the variability present in the data and also depends on a combination of the task to be performed and individual preference.

To ease the way for the illustrations to follow, compute and store each of the above-listed statistics in appropriately named variables.

```
y.hat<-fitted.values(Multiple.mod)

e<-residuals(Multiple.mod)

s<-Multiple.sum$sigma

#Compute standardized residuals

r<-e/s

#Extract leverages

h<-hatvalues(Multiple.mod)

#Compute studentized residuals

stud<-e/(s*sqrt(1-h))

#Extract studentized deleted residuals

stud.del<-rstudent(Multiple.mod)
```

Note that the functions `studres` located in the package `MASS` and `rstudent` in package `stats` obtain studentized *deleted* residuals, rather than studentized residuals. Similarly, the functions `stdres` in package `MASS` and `rstandard` in package `stats` compute standardized *deleted* residuals.

Additional *deletion statistics* that can play a role in the influence analysis of flagged outliers include *DFFITS*, *Cook's distances*, and *DFBETAS*. Brief descriptions along with formulas for computing these statistics are provided in the following sections.

Once again, depending on preference, numerical tests can be used to supplement graphical diagnostics. The methods applied for simple regression models apply here; however, the difficulty of working in dimensions higher than two increase the reliance on numerical methods in the analysis of outliers.

8.4.1   The constant variance assumption

Although the plots shown in Figure 8.3
serve mainly to assess the constant variances assumption, recall that they also serve to flag other potential issues such as the presence of outliers, the absence of important variables, or possible issues of fit. Ideal plots have all plotted points randomly scattered and approximately symmetrically spread about the horizontal axis.

Only the studentized residuals are used here; however, it should be remembered that any one of the three forms of (undeleted) residuals provide

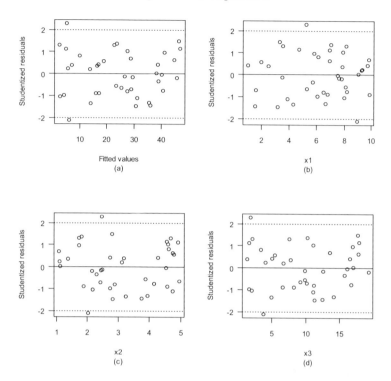

FIGURE 8.3: Plots of studentized residuals against the fitted values and explanatory variables.

equivalent information, but in potentially varying levels of clarity. The code used to produce Figure 8.3 is shown below.[2]

```
plot(y.hat,stud,xlab="Fitted values",las=1,
        ylab="Studentized residuals",sub="(a)")
abline(h=c(-2,0,2),lty=c(3,1,3))
attach(MultipleReg)
plot(x1,stud,xlab="x1",las=1,
        ylab="Studentized residuals",sub="(b)")
abline(h=c(-2,0,2),lty=c(3,1,3))
plot(x2,stud,xlab="x2",las=1,
        ylab="Studentized residuals",sub="(c)")
```

[2]Note that the functions `win.graph` and `par(mfrow=c(2,2))` were used to prepare the graphics window.

```
abline(h=c(-2,0,2),lty=c(3,1,3))
plot(x3,stud,xlab="x3",las=1,
        ylab="Studentized residuals",sub="(d)")
abline(h=c(-2,0,2),lty=c(3,1,3))
detach(MultipleReg)
```

The included horizontal lines are simply for reference and do not represent cutoff values. Note the argument `las` that is included in each of the `plot` functions. This instructs R on how to orient tick mark labels along the axes; `las=1` gets all these labels to be placed horizontally.

With respect to testing the constant variance assumption, the options are as for simple regression. If it is known that the error terms are approximately normally distributed, the F-test for the difference in variances between two populations can be used. If the normality assumption is in question, the Brown–Forsyth test should be preferred.

8.4.1.1 F-test for two population variances

For the given data and fitted model, the plots in Figure 8.3 really do not indicate a violation of the constant variances assumption. However, for the purpose of illustration, consider using the levels of `x1` in `MultipleReg` to rank the residuals for testing, and partition the data at `x1 = 6` (see Figure 8.3.b). Then,

```
Group<-with(MultipleReg,list("A"=cbind(e,x1)[x1<=6,1],
                             "B"=cbind(e,x1)[x1>6,1]))
var.test(Group$A,Group$B)
```

produces

```
            F test to compare two variances
data: Group$A and Group$B
F = 1.6214, num df = 15, denom df = 23, p-value = 0.2884
alternative hypothesis: true ratio of variances is not
equal to 1
95 percent confidence interval:
 0.6573976 4.3986183
sample estimates:
ratio of variances
            1.621439
```

Note the different approach used to obtain the two groups of residuals, as compared to that used for the simple regression case. The `list` function does not require equal lengths for its contents as does the function `data.frame`.

8.4.1.2 The Brown–Forsyth test

The earlier written Brown–Forsyth F-test function, bf.Ftest, can be used here. The function is first loaded in the current workspace using a command of the form

```
source("Z:/Docs/RCompanion/Chapter6/Functions/regBFFest.R")
```

Recall that the function call requires the following arguments: e, the residuals; x, the explanatory variable to be used for sorting the residuals; and cutoff, the value of x at which e is partitioned into two groups.

```
> with(MultipleReg,bf.Ftest(e,x1,6))

 Brown-Forsyth test using ranking by x1 split at x1 = 6

data: e

F = 1.9059, df1 = 1, df2 = 38, p-value = 0.1755

alternative hypothesis: Variances are unequal
```

The function bf.ttest produces equivalent results and the function hov from package HH can also be convinced to perform the same task.

8.4.2 The normality assumption

This may be accomplished by looking at a QQ normal probability plot of any one of the unstandardized, standardized, or studentized residuals. Figure 8.4, was obtained using the code

```
qqnorm(stud,main=NULL);abline(a=0,b=1)
```

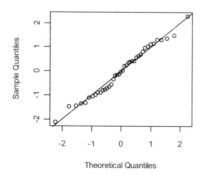

FIGURE 8.4: QQ normal probability plot of the studentized residuals along with the line $y = x$ for reference.

Observe, by setting `main=NULL`, the default figure title does not appear; not specifying an `lty` results in a default solid line. For the fitted model in question, the QQ plot of the standardized residuals looks very much like Figure 8.4.

The two tests of normality encountered earlier work here, too.

8.4.2.1 QQ normal probability correlation coefficient test

As for simple regression, first load the function `qq.cortest` using a command of the form,

```
source("z:/Docs/RCompanion/Chapter6/Functions/qqcortest.R")
```

then the test can be performed at $\alpha = .05$ on the earlier computed residuals, e, using

```
> qq.cortest(e,0.05)
  QQ Normal Probability Corr. Coeff. Test, alpha = 0.05
data: e, n = 40
RQ = 0.9916, RCrit = 0.972
alternative hypothesis: If RCrit > RQ, Normality assumption
is invalid
```

Since $r_Q > r_{\text{Crit}}$, the null hypothesis that the error terms are normally distributed is not rejected.

8.4.2.2 Shapiro–Wilk test

As before

```
> shapiro.test(stud)
    Shapiro-Wilk normality test
data: stud
W = 0.9815, p-value = 0.744
```

also indicating that there is not enough evidence to reject the assumption of normality (for $\alpha < p$).

8.4.3 The independence assumption

As with simple regression, any form of sequence plot of the residuals or a plot of the pairs $(\hat{\varepsilon}_i, \hat{\varepsilon}_{i+1})$ is useful in detecting a violation of the independence assumption. Plots of the residuals against time or spatial variables can be used to identify the possible presence of time or spatial dependence. Indications of first order autocorrelation are as for simple regression models.

In the case of time-series data, the Durbin–Watson test function, `dwtest`, can be used as described in Chapter 6 to test for the presence of first order autocorrelation.

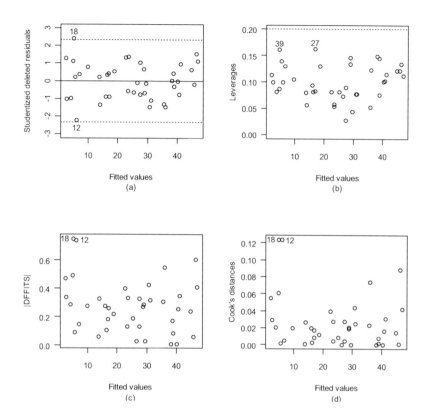

FIGURE 8.5: Plots for outlier diagnostics including studentized deleted residuals, leverages, DFFITS, and Cook's distances.

8.4.4 The presence and influence of outliers

Figures 8.5 and 8.6 provide means for graphical assessments of the measures presented below.[3]

8.4.4.1 Outlying observed responses

Figure 8.5.a, of the studentized deleted residuals, includes Bonferroni cutoff values, $\pm t(\alpha/(2m), n - p - 2)$ for the 5% most extreme values ($m = 2$). The code to obtain this plot is

```
m<-ceiling(.05*length(y.hat))
plot(stud.del~y.hat,xlab="Fitted values",sub="(a)",
```

---

[3]Note that the functions `win.graph` and `par` along with `mfrow=c(4,4)` were used to format the graphics window.

```
        ylab="Studentized deleted residuals",ylim=c(-3,3))
    cutoff<-qt(0.05/(2*m),length(stud.del)-3-2,lower.tail=F)
    abline(h=c(-cutoff,0,cutoff),lty=c(3,1,3))
    identify(y.hat,stud.del)
```

If only the outer 5% cases are tested ($m = 2$), then Case 18 is flagged as an outlier.

The function `outlierTest` from package `car` can also be used here; however, it must be remembered that this function looks at the indicated (`n.max`) most extreme cases with a Bonferroni adjustment using the whole sample size, $m = 40$ in the case of the current model. In using this test ($m = n$), if none of the studentized deleted residuals are flagged as outliers, then this test indicates there are none.

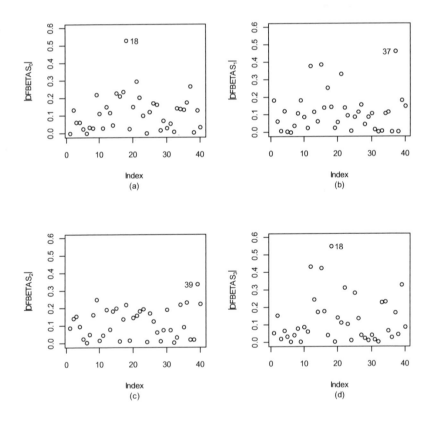

FIGURE 8.6: Index plots of DFBETAS to determine the influence of outlying cases on regression parameter estimates. Notice subscripts appearing in the vertical axis labels.

8.4.4.2 Outliers in the domain space

While experimental studies provide researchers with quite a bit of control on which values to use for the explanatory variables, observational studies might result in points in the *domain space* (x-value points) that could be considered outliers in relation to all other points. In simple regression, the task of identifying a potential outlying x-value is fairly simple: a boxplot does the job.

In multiple regression, the task is made difficult by the inclusion of additional dimensions. Leverages permit one way to flag such outlying points by providing a measure of the distance any given point $(x_{i1}, x_{i2}, \ldots, x_{ip})$ lies from the *centroid*, $(\bar{x}_1, \bar{x}_2, \ldots, \bar{x}_p)$, of the *domain space*. In Figure 8.5.b, the cutoff value of $2\bar{h}$ is used to test for outlying cases, where $\bar{h}$ represents the mean leverage value. The code used to obtain these plots is as for Figure 8.5.a, except that h is used in place of stud.del.

8.4.4.3 Influence of outliers on corresponding fitted values

The influence an individual $(p+1)$-tuple, say $(y_i, x_{i1}, x_{i2}, \ldots, x_{ip})$, has on the corresponding fitted value $\hat{y}_i$ can be examined by looking at the magnitude of $DFFITS_i$, which is computed using

$$DFFITS_i = \hat{d}_i^* \sqrt{\frac{h_{ii}}{1 - h_{ii}}}.$$

The function to compute $DFFITS$ is

```
DFFITS<-dffits(Multiple.mod)
```

Two commonly accepted yardsticks for $DFFITS$, see for example [45, p. 401], are:

1. If n is small to medium in size, then $|DFFITS_i| > 1$ suggests that the ith case may be influential.

2. If n is large, then $|DFFITS_i| > 2\sqrt{(p+1)/n}$ suggests that the ith case may be influential.

Plots equivalent to those used for outlier flagging can be used here; see Figure 8.5.c. The code used is a slight modification of before, with abs(DFFITS) being used in place of stud.del and h.

Observe that the cutoff line of 1 is not needed in these plots. It should be noted that although none of the plotted points lie anywhere near 1, Cases 12 and 18 are relatively more influential on their corresponding fitted values. These two cases also stood out in the plot of the studentized deleted residuals in Figure 8.5.a, but not in the plot of the leverages in Figure 8.5.b.

The big question in looking at $DFFITS$ is, what defines n as being small, medium, or large? If the yardstick for large samples is used, Cases 12 and 18 are identified as being influential.

8.4.4.4   Influence of outliers on all fitted values

The effect that an individual $(p+1)$-tuple, say $(y_i, x_{i1}, x_{i2}, \ldots, x_{ip})$, has on every fitted value $\hat{y}_k$, $k = 1, 2, \ldots, n$, can be examined by looking at the magnitude of Cook's distance D_i, which is computed using

$$D_i = \frac{\hat{\varepsilon}_i^2}{(p+1)\, s^2} \left[\frac{h_{ii}}{(1-h_{ii})^2} \right].$$

The function to compute Cook's distances is

```
Cook<-cooks.distance(Multiple.mod)
```

There is some debate on what constitutes a "safe" yardstick when using Cook's distances. One suggestion, see for example [45, p. 403], is that the ith case be considered influential if Cook's distance satisfies

$$D_i > F(0.5, p+1, n-p-1).$$

However, the suggestion that appears fairly consistently is that Cook's distances should be used in concert with all other influence measures *and* observations made from graphs. The code for Figure 8.5.d is as before with Cook being used in the vertical axis. The cutoff line can be inserted via the function abline and qf.

```
abline(h=qf(0.5,4,36),lty=3)
```

Once again, Cases 12 and 18 stand out, suggesting they are relatively more influential than other cases. However, both plotted points also lie well below the cutoff value.

8.4.4.5   Influence of outliers on parameter estimates

Finally, the effect that an individual outlying case, say $(y_i, x_{i1}, x_{i2}, \ldots, x_{ip})$, has on the kth parameter estimate, $\hat{\beta}_k$, can be examined by looking at the magnitude of $DFBETAS_{k(i)}$.

The computation procedure for these statistics is as follows: If $\hat{\beta}_{k(i)}$ is the kth regression coefficient for the ith *deleted model* (i.e., the model obtained by removing the ith case), then for $k = 0, 1, 2, \ldots, p$,

$$DFBETAS_{k(i)} = \frac{\hat{\beta}_k - \hat{\beta}_{k(i)}}{\sqrt{s_{(i)}^2 \, c_{kk}}},$$

where c_{kk} is the kth diagonal entry of $(\mathbf{X'X})^{-1}$ and $s_{(i)}^2$ is the mean square error of the ith deleted model. This last collection of deletion statistics would be a bear to obtain. Fortunately, R has a function that computes each of these deletion statistics.

```
DFBETAS<-dfbetas(Multiple.mod)
```

Yardsticks available for $DFBETAS$ are as follows, see for example [45, pp. 404–405].

1. If n is small to medium in size, then $|DFBETAS_{k(i)}| > 1$ suggests that the ith case may be influential on $\hat{\beta}_k$.

2. If n is large, then $|DFBETAS_{k(i)}| > 2/\sqrt{n}$ suggests that the ith case may be influential on $\hat{\beta}_k$.

Since there are $p+1$ parameter estimates, $p+1$ plots are needed; four for the model in question. Here, consider looking at only the *index plots* of *absolute values* of the DFBETAS. To ensure some consistency in the vertical range of the plots, a quick inspection of the contents of $DFBETAS$ indicates the largest magnitude is less than 0.6,

```
> max(abs(DFBETAS))
[1] 0.5513183
```

Figure 8.6 was obtained using the code

```
plot(abs(DFBETAS[,1]),ylim=c(0,0.6),sub="(a)",
        ylab=expression(abs(DFBETAS[0])))
identify(abs(DFBETAS[,1]))
plot(abs(DFBETAS[,2]),ylim=c(0,0.6),sub="(b)",
        ylab=expression(abs(DFBETAS[1])))
identify(abs(DFBETAS[,2]))
plot(abs(DFBETAS[,3]),ylim=c(0,0.6),sub="(c)",
        ylab=expression(abs(DFBETAS[2])))
identify(abs(DFBETAS[,3]))
plot(abs(DFBETAS[,4]),ylim=c(0,0.6),sub="(d)",
        ylab=expression(abs(DFBETAS[3])))
identify(abs(DFBETAS[,4]))
```

None of the extreme cases in Figure 8.6 are larger than 1, so one may conclude that there are no influential cases. However, Case 18 is one that does stand out. Keep in mind that opinions also vary on yardsticks used in analyzing DFBETAS.

8.4.4.6   Quick influence analysis

A quick way to assess the influence of individual cases is to make use of the function `influence.measures` from package `stats`, along with some selective searches.

To identify influential cases, one could browse through the output of

```
influence.measures(Multiple.mod)
```

However, even with the medium sized dataset in use, this can get tiresome. A short program can be prepared to sift through the output and extract only the cases flagged as being influential. For example,

```
#Get all results
All<-influence.measures(Multiple.mod)
#Separate test results and measures
tests<-data.frame(All[2]);measures<-data.frame(All[1])
#Now sift out the cases that are flagged influential
#First initialize the starter counter
j<-0
#Loop through all 40 cases
for (i in 1:40){
    #Set the test result condition
    is.inf<-FALSE
    #Loop through all 8 columns of All
    for (k in 1:8){
        #Check ith case for influence across 8 columns
        is.inf<-isTRUE((tests[i,k]==TRUE)|is.inf)}
    #If the ith case is influential by one or more measures
    if (is.inf==TRUE){
    #Store the case number
    if (j==0){inf.cases<-i;j<-1
        } else {inf.cases<-c(inf.cases,i)}}}
#Now pull out only the influential cases
inf.details<-data.frame(measures[inf.cases,])
#Give the columns names
names(inf.details)<-c("dfb0","dfb1","dfb2","dfb3",
                    "dffits","cov.r","cook.d","hat")
#Output the findings
round(inf.details,3)
```

produces

```
    dfb0    dfb1    dfb2    dfb3   dffits  cov.r  cook.d    hat
18 0.529  -0.148  -0.221  -0.551    0.75  0.654   0.124  0.086
```

For the current example, only 18 shows up. To find out which particular influence measures flagged Case 18 as being influential, execute

```
names(tests)<-names(inf.details)
tests[inf.cases,]
```

to get

```
    dfb0    dfb1    dfb2    dfb3 dffits cov.r cook.d    hat
18 FALSE   FALSE   FALSE   FALSE  FALSE  TRUE  FALSE  FALSE
```

So, Case 18 is flagged because of the covariance ratio value. Some discussion might be appropriate here.

The quantity cov.r[i] is the ratio of the determinant of the variance-covariance matrix of the ith deleted model and the determinant of the variance-covariance matrix of the undeleted model. This provides a measure of the comparitive "size" of the confidence region for the regression parameters of the ith deleted model in relation to that of the undeleted model. A computational formula for this statistic is

$$COVRATIO_i = 1 \left/ \left[(1 - h_{ii}) \left(\frac{(n - p - 2) + \hat{d}_i^{*2}}{n - p - 1} \right)^{p+1} \right] \right. ,$$

and an accepted yardstick is to consider the ith case worth investigating for influence if

$$|COVRATIO_i - 1| \geq \frac{3(p + 1)}{n} .$$

In the example under consideration, the covariance ratio for Case 18 lies above the cut-off value.

Notice that $COVRATIO_i$ is large when h_{ii} is large (close to 1) or when $\hat{d}_i^{*2}$ is small (close to zero). There is the possibility that an observation results in both a large h_{ii} and a large $\hat{d}_i^{*2}$, which would be a point of interest. However, the value of $COVRATIO_i$ would most likely remain within the recommended bounds because of the "neutralizing" effect each term has on the other. For this reason it is suggested that $COVRATIO_i$ is not (always) a reliable measure of influence. See, for example, [5, pp. 22–23], [50, p. 216], and [53, pp. 293–298] for further details on the subject of influence analysis.

8.5   Estimating Regression Parameters

As with simple regression, one-at-a-time intervals and simultaneous intervals can be obtained.

8.5.1   One-at-a-time t-intervals

The formula for an interval of this type remains the same,

$$\hat{\beta}_j - t(\alpha/2, n - p - 1)\, s_{\hat{\beta}_j} < \beta_j < \hat{\beta}_j + t(\alpha/2, n - p - 1)\, s_{\hat{\beta}_j}.$$

The quantities needed are $\hat{\beta}_j$ and $s_{\hat{\beta}_j}$, which are available in `Multiple.sum`, and $t(\alpha/2, n - p - 1)$, which can be computed using the function `qt`. Recall that the relevant statistics for β_0 are stored in the first (numeric) row of the `coefficients` object in `Multiple.sum`, hence relevant statistics for β_j are stored in the $(j+1)^{\text{st}}$ row of `coefficients`. As demonstrated in the previous chapter, code can be written to obtain such intervals using the above formula. The more efficient way is to make use of the R function `confint`. Thus,

```
> round(confint(Multiple.mod,parm="x2",level=1-.05),3)
      2.5 %  97.5 %
x2    0.596   1.113
```

Recall that `level` represents the confidence level and that the default value for this argument is 95%. Also, if `parm` is left out, intervals for all parameters are calculated.

8.5.2   Simultaneous t-intervals

The general formula for such intervals is

$$\hat{\beta}_j - t(\alpha/(2m), n - p - 1)\, s_{\hat{\beta}_j} < \beta_j < \hat{\beta}_j + t(\alpha/(2m), n - p - 1)\, s_{\hat{\beta}_j},$$

where m represents the number of parameters being estimated and j the parameter subscript.

Once again, code may be written to perform Bonferroni's procedure; however, the function `confint` works well here too. The key is to be sure to make the Bonferroni adjustment to the confidence `level` passed into the function. Consider obtaining simultaneous 95% intervals for β_0 and β_1; that is, $m = 2$. The adjustment is passed into `confint` through the argument `level`. Then

```
round(confint(Multiple.mod,
    parm=c("(Intercept)","x1"),level=1-.05/2),3)
```

produces

```
                 1.25 %   98.75 %
  (Intercept)    1.379     4.504
  x1            -1.057    -0.763
```

Notice that the significance level $\alpha = 0.05$ is distributed evenly between the two intervals resulting in individual confidence levels of 97.5%. As long as the number of parameters being estimated is not too large, this procedure works well.

8.5.3   Simultaneous F-intervals

When the number of simultaneous intervals sought is large, Scheffe's procedure might be preferred. The general formula for the Scheffe intervals for all parameters in a model is

$$\hat{\beta}_j - \sqrt{(p+1)\,F\,(\alpha, p+1, n-p-1)}\,s_{\hat{\beta}_j}$$
$$< \beta_j <$$
$$\hat{\beta}_j + \sqrt{(p+1)\,F\,(\alpha, p+1, n-p-1)}\,s_{\hat{\beta}_j},$$

and the quantities needed are $\hat{\beta}_j$ and $s_{\hat{\beta}_j}$, which are available in `Multiple.sum` and $F\,(\alpha, p+1, n-p-1)$, which can be computed using the function `qf`. There does not appear to be an R function for computing Scheffe intervals; however, the following code does the job:

```
#Attach summary object for convenience
attach(Multiple.sum)
#Initialize various objects
alpha<-0.05;b<-numeric(4);sb<-numeric(4)
lb<-numeric(4);rb<-numeric(4)
#Calculate "critical" value
cv<-sqrt(df[1]*qf(alpha,df[1],df[2],lower.tail=F))
#Loop through interval computations
for (j in 1:4){
   b[j]<-coefficients[j,1]; sb[j]<-coefficients[j,2]
   lb[j]<-b[j]-cv*sb[j]; rb[j]<-b[j]+cv*sb[j]}
#Store results and name rows appropriately
intervals<-data.frame("Estimate"=b,
```

```
"Error"=sb,"Lower"=lb,"Upper"=rb)
row.names(intervals)<-c("Intercept","x1","x2","x3")
#Detach summary object to be safe
detach(Multiple.sum)
#Take a look at the results
round(intervals,3)
```

producing

	Estimate	Error	Lower	Upper
Intercept	2.941	0.668	0.773	5.109
x1	-0.910	0.063	-1.114	-0.706
x2	0.854	0.127	0.441	1.268
x3	2.492	0.027	2.403	2.581

There are occasions when joint confidence regions of regression parameters might be desired. The last section of this chapter provides an illustration of plotting such regions.

8.6   Confidence Intervals for the Mean Response

As with simple regression, there may be a need to obtain a single interval, or one might wish to obtain simultaneous intervals.

8.6.1   One-at-a-time t-intervals

Let $\mathbf{x}' = (1, x_1, x_2, \ldots, x_p)$, where $(x_1, x_2, \ldots, x_p)$ denotes a given p-tuple for the explanatory variables. Remember that R will store $\mathbf{x}$ as a column vector. Then, in algebraic form, the true response at the point $(x_1, x_2, \ldots, x_p)$ is given by

$$y = \beta + \beta_1 x_1 + \beta_2 x_2 + \cdots + \beta_p x_p + \varepsilon,$$

and the true mean response is

$$E(y) = \beta + \beta_1 x_1 + \beta_2 x_2 + \cdots + \beta_p x_p.$$

The fitted value at $(x_1, x_2, \ldots, x_p)$ is given by

$$\hat{y} = \hat{\beta}_0 + \hat{\beta}_1 x_1 + \hat{\beta}_2 x_2 + \cdots + \hat{\beta}_p x_p.$$

The estimated variance of $\hat{y}$ is given by

$$s_{\hat{y}}^2 = \mathbf{x}'(\mathbf{X}'\mathbf{X})^{-1}\mathbf{x}\, s^2,$$

and a one-at-a-time confidence interval for the mean response at $(x_1, x_2, \ldots, x_p)$ is computed using

$$\hat{y} - t(\alpha/2, n-p-1)\, s_{\hat{y}} < \mathrm{E}(y) < \hat{y} + t(\alpha/2, n-p-1)\, s_{\hat{y}}.$$

Thus, items needed from `Multiple.sum` for the computations include: the matrix $(\mathbf{X}'\mathbf{X})^{-1}$, which is stored as `cov.unscaled`; s, which is stored as `sigma`; and $\hat{\boldsymbol{\beta}}$, which is stored in the first column of `coefficients`. As before, $t(\alpha/2, n-p-1)$ is computed using the function `qt`.

Once again, a short program can be written to produce this confidence interval, but it is more efficient to use the function `predict` with the argument `interval="confidence"`.

Consider obtaining a 95% confidence interval for the mean response at $(x_1, x_2, x_3) = (7.36, 0.41, 18.49)$. Then,

```
#Store the point in a data frame
x<-data.frame(x1=7.36,x2=.41,x3=18.49)
#Obtain the bounds
predict(Multiple.mod,x,interval="confidence")
```

produces

```
        fit      lwr      upr
1 42.6667 41.75816 43.57524
```

where `fit` is the fitted value at the given x.

8.6.2   Simultaneous t-intervals

Suppose m simultaneous confidence intervals for the mean response are desired. For $k = 1, 2, \ldots, m$, let the row vectors $\mathbf{x}'_k = (1, x_{k1}, x_{k2}, \ldots, x_{kp})$ represent the m levels (or points of interest) of the explanatory variables, with each $(x_{k1}, x_{k2}, \ldots, x_{kp})$ being a given or observed p-tuple, then the fitted values and corresponding variances are

$$\hat{y}_k = \hat{\beta}_0 + \hat{\beta}_1 x_{k1} + \hat{\beta}_2 x_{k2} + \cdots + \hat{\beta}_p x_{kp} \quad \text{and} \quad s_{\hat{y}_k}^2 = \mathbf{x}'_k(\mathbf{X}'\mathbf{X})^{-1}\mathbf{x}_k\, s^2.$$

Bonferroni's procedure uses the formula

$$\hat{y}_k - t(\alpha/(2m), n-p-1)\, s_{\hat{y}_k} < \mathrm{E}(y_k) < \hat{y}_k + t(\alpha/(2m), n-p-1)\, s_{\hat{y}_k},$$

and can be performed using the `predict` function.

Suppose simultaneous t-intervals are desired at the points

$$(x_{11}, x_{12}, x_{13}) = (7.7, 2.3, 9.6), \qquad (x_{21}, x_{22}, x_{23}) = (8.5, 2.6, 13.1),$$
$$(x_{31}, x_{32}, x_{33}) = (1.5, 3.1, 3.4), \qquad (x_{41}, x_{42}, x_{43}) = (5.8, 2.0, 5.2).$$

First, create a data frame for the points of interest, note the number of simultaneous intervals desired, and choose α.

```
xm<-data.frame(cbind(x1=c(7.7,8.5,1.5,5.8),
                     x2=c(2.3,2.6,3.1,2.0),
                     x3=c(9.6,13.1,3.4,5.2)))
m<-dim(xm)[1]; alpha<-.05
```

Then, adjusting the significance level for m simultaneous intervals, enter

```
predict(Multiple.mod,xm,
        interval="confidence",level=1-alpha/m)
```

to get

```
     fit       lwr      upr
1 21.82069 21.31524 22.32615
2 30.07013 29.48887 30.65138
3 12.69659 11.76253 13.63064
4 12.32955 11.70166 12.95744
```

As is the case for simple regression, F-intervals might be preferred in cases where a much larger number of simultaneous intervals are desired.

8.6.3 Simultaneous F-intervals

Applying the *Working–Hotelling procedure* to multiple regression models is analogous to the case of simple regression models. Code to implement the formula

$$\hat{y}_k - \sqrt{(p+1)\,F\,(\alpha, p+1, n-p-1)}\,s_{\hat{y}_k}$$
$$< E(y_k) <$$
$$\hat{y}_k + \sqrt{(p+1)\,F\,(\alpha, p+1, n-p-1)}\,s_{\hat{y}_k},$$

can be prepared. Working with the same points

$$(x_{11}, x_{12}, x_{13}) = (7.7, 2.3, 9.6), \qquad (x_{21}, x_{22}, x_{23}) = (8.5, 2.6, 13.1),$$
$$(x_{31}, x_{32}, x_{33}) = (1.5, 3.1, 3.4), \qquad (x_{41}, x_{42}, x_{43}) = (5.8, 2.0, 5.2);$$

the code

```
#Store the points in a matrix
xm<-cbind(rep(1,4),c(7.7,8.5,1.5,5.8),
c(2.3,2.6,3.1,2.0),c(9.6,13.1,3.4,5.2))
#Initialize the various variables
m<-dim(xm)[1]; fit<-numeric(m)
lwr<-numeric(m); upr<-numeric(m); a<-0.05
#For easy access
attach(Multiple.sum)
#Extract the various statistics needed
b<-coefficients[,1]; s<-sigma; XtX.inv<-cov.unscaled
#Compute the "critical" value
cv<-sqrt(df[1]*qf(a,df[1],df[2],lower.tail=FALSE))
detach(Multiple.sum)
#Loop through the four points and compute
for (k in 1:m){
    #the fitted value
    fit[k]<-sum(xm[k,]*b)
    #the standard error, note use of the operator %*%
    s.er<-s*sqrt(t(xm[k,])%*%XtX.inv%*%xm[k,])
    #Lower and upper bounds
    lwr[k]<-fit[k]-cv*p.er; upr[k]<-fit[k]+cv*p.er}
#Store the results and output for viewing
results<-data.frame(fit,lwr,upr)
round(results,4)
```

produces

```
      fit     lwr     upr
1 21.8207 21.1968 22.4446
2 30.0701 29.3527 30.7876
3 12.6966 11.5436 13.8495
4 12.3296 11.5545 13.1046
```

As shown in the last section of this chapter, this procedure also provides a way of obtaining a confidence region for the mean response for the entire regression "surface."

To avoid possible complications with tasks in the next section, remove all objects from the workspace *except* Multiple.mod, Multiple.sum, and MultipleReg.

8.7 Prediction Intervals for New Observations

As before, the one computational difference between obtaining confidence intervals for the mean response and prediction intervals for new responses at a given point $(x_1, x_2, \ldots, x_n)$ is that the estimated variance used for prediction intervals is

$$s^2_{\hat{y}_{new}} = \left(1 + \mathbf{x}'(\mathbf{X}'\mathbf{X})^{-1}\mathbf{x}\right) s^2,$$

where $\mathbf{x}$ is as defined earlier. This is used in place of $s_{\hat{y}_0}$ (or $s_{\hat{y}_k}$) for all code presented for interval estimates for the mean response. The function `predict` can be used to obtain prediction t-intervals by specifying the argument `interval="prediction"`.

8.7.1 t-interval for a single new response

With the above-mentioned alteration, the formula becomes

$$\hat{y}_{new} - t(\alpha/2, n - p - 1)\, s_{\hat{y}_{new}} < y_{new} < \hat{y}_{new} + t(\alpha/2, n - p - 1)\, s_{\hat{y}_{new}},$$

with

$$\hat{y}_{new} = \hat{\beta}_0 + \hat{\beta}_1 x_1 + \hat{\beta}_2 x_2 + \cdots + \hat{\beta}_p x_p.$$

Consider obtaining a 95% prediction interval for the response to $(x_1, x_2, x_3) = (7.36, 1.41, 18.49)$ using the function `predict`. Then,

```
x<-data.frame(x1=7.36,x2=1.41,x3=18.49)
predict(Multiple.mod,x,interval="prediction")
```

gives

```
        fit      lwr      upr
1 43.52104 41.47249 45.5696
```

If so desired, this result can be duplicated with code from scratch.

8.7.2 Simultaneous t-intervals

The process is as before, with the exception that $s^2_{\hat{y}_{new}(k)}$ is used instead of $s^2_{\hat{y}_k}$. Suppose simultaneous intervals for m new observations are desired and suppose

$$
\begin{aligned}
\mathbf{x}'_1 &= (1, x_{11}, x_{12}, \ldots, x_{1p}); \\
\mathbf{x}'_2 &= (1, x_{21}, x_{22}, \ldots, x_{2p}); \\
&\vdots \\
\mathbf{x}'_m &= (1, x_{m1}, x_{m2}, \ldots, x_{mp}).
\end{aligned}
$$

Then

$$\hat{y}_{\text{new}(k)} = \hat{\beta}_0 + \hat{\beta}_1 x_{k1} + \hat{\beta}_2 x_{k2} + \cdots + \hat{\beta}_p x_{kp},$$

$$s^2_{\hat{y}_{\text{new}(k)}} = \left(1 + \mathbf{x}'_k (\mathbf{X}'\mathbf{X})^{-1} \mathbf{x}_k\right) s^2$$

and, at a joint significance level of α, the Bonferroini intervals are given by

$$\hat{y}_{\text{new}(k)} - t(\alpha/(2m), n - p - 1) \, s_{\hat{y}_{\text{new}(k)}}$$

$$< y_{\text{new}(k)} <$$

$$\hat{y}_{\text{new}(k)} + t(\alpha/(2m), n - p - 1) \, s_{\hat{y}_{\text{new}(k)}}.$$

Suppose for the model in question simultaneous t-intervals are desired at the points

$$\begin{aligned}
(x_{11}, x_{12}, x_{13}) &= (7.7, 2.3, 9.6), & (x_{21}, x_{22}, x_{23}) &= (8.5, 2.6, 13.1), \\
(x_{31}, x_{32}, x_{33}) &= (1.5, 3.1, 3.4), & (x_{41}, x_{42}, x_{43}) &= (5.8, 2.0, 5.2).
\end{aligned}$$

Using the function `predict`, the process is as for Bonferroni confidence intervals of the mean response with the exception that `interval="prediction"`. The code

```
xm<-data.frame(x1=c(7.7,8.5,1.5,5.8),
                x2=c(2.3,2.6,3.1,2.0),
                x3=c(9.6,13.1,3.4,5.2))
m<-4; alpha<-.05
predict(Multiple.mod,xm,
        interval="prediction",level=1-alpha/m)
```

gives

```
      fit        lwr        upr
1 21.82069  19.286906  24.35448
2 30.07013  27.520137  32.62012
3 12.69659  10.043844  15.34933
4 12.32955   9.768529  14.89057
```

As with confidence intervals for the mean response, F-intervals can be obtained for prediction intervals.

8.7.3   Simultaneous F-intervals

Using the same points from the previous illustration, the code used to obtain F-intervals for the mean response can be altered to work for computing

$$\hat{y}_{\text{new}(k)} - \sqrt{m\,F\left(\alpha, m, n - p - 1\right)}\, s_{\hat{y}_{\text{new}(k)}}$$

$$< y_{\text{new}(k)} <$$

$$\hat{y}_{\text{new}(k)} + \sqrt{m\,F\left(\alpha, m, n - p - 1\right)}\, s_{\hat{y}_{\text{new}(k)}}$$

as follows, with changes indicated by comments:

```
xm<-cbind(rep(1,4),c(7.7,8.5,1.5,5.8),
                   c(2.3,2.6,3.1,2.0),
                   c(9.6,13.1,3.4,5.2))
m<-dim(xm)[1]; fit<-numeric(m)
lwr<-numeric(m); upr<-numeric(m); a<-0.05
attach(Multiple.sum)
b<-coefficients[,1]; s<-sigma; XtX.inv<-cov.unscaled
#A different formula is needed here
cv<-sqrt(m*qf(a,m,df[2],lower.tail=F))
detach(Multiple.sum)
for (k in 1:m){
    fit[k]<-sum(xm[k,]*b)
    #A different formula is need here
    s.er<-s*sqrt(1+t(xm[k,])%*%XtX.inv%*%xm[k,])
    lwr[k]<-fit[k]-cv*s.er; upr[k]<-fit[k]+cv*s.er}
results<-data.frame(fit,lwr,upr)
round(results,4)
```

The results are

```
      fit      lwr      upr
1 21.8207 18.6931 24.9482
2 30.0701 26.9226 33.2177
3 12.6966  9.4222 15.9710
4 12.3296  9.1684 15.4907
```

Prediction regions can also be computed; see Section 8.8.4 for a brief discussion on one way to do this.

8.8 For the Curious

Traditional to some courses on multiple regression is an introduction to (at least a flavor of) the matrix approach to regression analysis; see, for example, [45, Chapters 5 and 6]. The task of demonstrating this approach is made quite painless using R's computational capabilities. Here, an illustration is provided by first fitting the dataset `MultipleReg` to the earlier proposed model, and then performing the goodness of fit F-test on the fitted model. Also given are examples of code used to obtain confidence regions associated with regression parameters, as well as the mean response.

Clear the workspace of all objects except `MultipleReg`, `Multiple.mod` and `Multiple.sum` before continuing.

8.8.1 Fitting and testing a model from scratch

Recall that it was stated that the model in matrix form is

$$\mathbf{y} = \mathbf{X}\boldsymbol{\beta} + \varepsilon$$

where the *design matrix*, $\mathbf{X}$, has *full column rank*; the *response vector* $\mathbf{y}$ is a solution of $\mathbf{y} = \mathbf{X}\boldsymbol{\beta} + \varepsilon$; and the entries of the *error vector*, ε, satisfy the earlier stated conditions. Through some probability theory and calculus, it can be shown that the unique least squares solution, $\hat{\boldsymbol{\beta}}$, to the above equation is the solution of what is called the *normal equation* for the model,

$$\mathbf{X}'\mathbf{X}\hat{\boldsymbol{\beta}} = \mathbf{X}'\mathbf{y}.$$

The condition that $\mathbf{X}$ has full column rank permits the solution

$$\hat{\boldsymbol{\beta}} = (\mathbf{X}'\mathbf{X})^{-1}\mathbf{X}'\mathbf{y}.$$

Now, to work from scratch, the first task would be to construct the vector $\mathbf{y}$ and the matrix $\mathbf{X}$ from the data. This is simple enough using

```
y<-as.vector(MultipleReg$y); n<-length(y)
X<-with(MultipleReg,as.matrix(cbind(rep(1,n),x1,x2,x3)))
```

Out of curiousity, execute the command

```
model.matrix(Multiple.mod)
```

and compare what you see with X. Now, compute $\mathbf{X}'\mathbf{X}$ and $(\mathbf{X}'\mathbf{X})^{-1}$

```
XtX<-t(X)%*%X; XtX.inv<-solve(XtX)
```

and then execute

```
Multiple.sum$cov.unscaled
```

and compare what you see with XtX.inv. Finally, calculate $(\mathbf{X}'\mathbf{X})^{-1}\mathbf{X}'\mathbf{y}$

```
b<-XtX.inv%*%(t(X)%*%y)
```

and compare the results with

```
Multiple.mod$coefficients
```

The vectors of fitted values and residuals can then be computed using

```
y.hat<-X%*%b; e<-y-y.hat
```

The residual sum of squares, the mean square error, and the residual standard error can be computed using

```
SSE<-sum(e^2); MSE<-SSE/(n-4); s<-sqrt(MSE)
```

The goodness of fit F-test looks at

$\text{H}_0 : y_i = \beta_0 + \varepsilon_i$ {Reduced model} vs.

$\text{H}_1 : y_i = \beta_0 + \beta_1 x_{i1} + \beta_2 x_{i2} + \beta_3 x_{i3} + \varepsilon_i$ {Full model}

which requires computing the test statistic

$$F^* = MSR/MSE.$$

What remains, then, is to go through the same process for the reduced model to compute the sum of squares of total variation, $SSTo$. Here is the code. Note that the design matrix for the reduced model is simply a column of 1s.

```
I<-as.matrix(rep(1,n))
b.red<-solve(t(I)%*%I)%*%(t(I)%*%y)
y.red<-I%*%b.red; e.red<-y-y.red
SSTo<-sum(e.red^2)
```

Then

```
SSR<-SSTo-SSE; MSR<-SSR/3
F.stat<-MSR/MSE
p.value<-pf(F.stat,3,n-4,lower.tail=F)
```

and the ANOVA table for the fitted model can be put together using

```
f.test<-data.frame(
    "df"=c(3,n-4,n-1),
    "SS"=c(SSR,SSE,SSTo),
    "MS"=c(MSR,MSE," "),
    "F.Stat"=c(F.stat," "," "),
    "p.value"=c(round(p.value,4)," "," "),
    row.names=c("Regression","Error","Total"))
```

Rounding as appropriate shows that `f.test` contains

	df	SS	MS	F.Stat	p.value
Regression	3	7603.72	2534.57	2842.70	0
Error	36	32.09	0.89		
Total	39	7635.82			

Compare this with results from `Multiple.sum`.

8.8.2 Joint confidence regions for regression parameters

Plots of pairwise confidence regions can be obtained via Sheffe's procedure. Suppose a joint region for β_j and β_k is desired. Let $\mathbf{A}$ be a $2 \times (p+1)$ matrix such that

$$\boldsymbol{\theta} = \mathbf{A}\boldsymbol{\beta} = \begin{pmatrix} \beta_j \\ \beta_k \end{pmatrix},$$

and let $\hat{\boldsymbol{\theta}} = \mathbf{A}\hat{\boldsymbol{\beta}}$ be the vector containing the estimates $\hat{\beta}_j$ and $\hat{\beta}_k$. Then, the ellipse described by the inequality

$$\left(\hat{\boldsymbol{\theta}} - \boldsymbol{\theta}\right)' \left[\mathbf{A} \left(\mathbf{X}'\mathbf{X}\right)^{-1} \mathbf{A}'\right]^{-1} \left(\hat{\boldsymbol{\theta}} - \boldsymbol{\theta}\right) \leq 2\,s^2\,F\left(\alpha, 2, n - p - 1\right)$$

describes the $(1-\alpha)\%$ joint confidence region for the parameters β_j and β_k in two-space. Writing a program to obtain this region is complicated enough to prompt a search for a readily available function to do the task. The function `ellipse` from package `ellipse` [52] produces ordered pairs that bound this region. The syntax for computing bounds for the $(1 - \alpha)\%$ joint confidence region for two parameters (say β_j and β_k) is

```
ellipse(model,which=c(j,k),level=(1-alpha))
```

The argument `which` is assigned values as shown above, or in a manner such as `which=c("xj","xk")`, the argument `"(intercept)"` is used if β_0 is one of the parameters. The `plot` function can then be used to plot the ellipsoid that bounds the computed region.

Consider a joint confidence region for β_1 and β_2 (keep in mind that, in R, the indexing for coefficients is $1, 2, \ldots, (p + 1)$ rather than $0, 1, \ldots, p$). First, get the bare-bones plot using

```
#Set parameter identifiers
j<-2;k<-3
#Load package ellipse
require(ellipse)
#Obtain region bounds and plot
region<-ellipse(Multiple.mod,which=c(j,k),level=.95)
plot(region,type="l",xlab="",main=NULL,ylab="",las=1,
    xlim=c(min(c(0,region[,1])),max(c(0,region[,1]))),
        ylim=c(min(c(0,region[,2])),max(c(0,region[,2]))))
detach(package:ellipse)
```

Then add on some bells and whistles to get Figure 8.7.

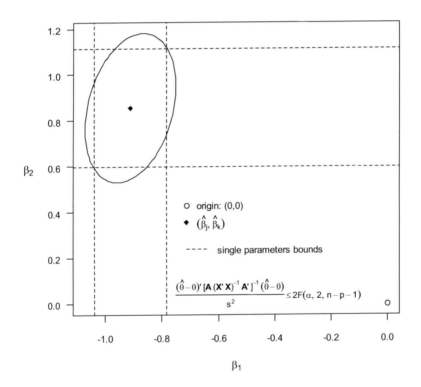

FIGURE 8.7: Plot of joint confidence region for β_1 and β_2 along with some unnecessary bells and whistles.

```
#Place horizontal axis label
mtext(bquote(beta[.(j-1)]),side=1,line=3,cex=1.15)
#Place vertical axis label
mtext(bquote(beta[.(k-1)]),side=2,line=3,cex=1.15,las=1)
points(0,0) #Plot the point (0,0)
```

#Plot the point $(\hat{\beta}_1, \hat{\beta}_2)$

```
points(coef(Multiple.mod)[j],coef(Multiple.mod)[k],pch=18)
```

#Plot lines representing bounds for β_1 and β_2

```
abline(v=confint(Multiple.mod)[j,],lty=2)
abline(h=confint(Multiple.mod)[k,],lty=2)
#Put in fancy-pants mathematical annotation
text(-.75,.05,
    expression(frac(paste((hat(bold(theta))-bold(theta)),
        "' [",bold(A)," (",bold(X),"' ",bold(X),
        ")"^-1," ",bold(A),"' ]"^-1," ",
        (hat(bold(theta))-bold(theta))),s^2)
        <=2*F(alpha,2,n-p-1)),cex=.9,adj=0)
```

#Put in a legend for plotted points $(0,0)$ and $(\hat{\beta}_1, \hat{\beta}_2)$

```
legend(-.75,.5,pch=c(1,18),box.lty=0,
    legend=c("origin: (0,0)",
        expression((list(hat(beta)[j],hat(beta)[k])))))
legend(-.75,.3,lty=2,box.lty=0,
        legend="single parameters bounds")
```

Included in this last set of code are examples of how to paste mathematical annotation on a plot, more on how to include legends, a way to automate x- and y-limits for the axes, and a way to automate axes labels.

8.8.3 Confidence regions for the mean response

In simple regression, this "surface" translates to a confidence band in the xy-plane. In multiple regression, things get a bit more complicated because of the increased number of dimensions. One may obtain a two-dimensional "slice" of a region by requiring that all but one of the explanatory variables be fixed (see Figure 8.8). Both Bonferroni's and the Working–Hotelling procedures can be used for this purpose.

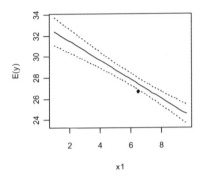

FIGURE 8.8: Bonferroni bands for the mean response with x_2 and x_3 held fixed.

To compute regions using Bonferroni's procedure, the code for simultaneous intervals can be altered to obtain the desired bounds on $E(y_i)$ corresponding to the p-tuple $(x_{i1}, x_{i2}, \ldots, x_{ip})$, for each $i = 1, 2, \ldots, n$. First, let $\mathbf{x}'_i = (1, x_{i1}, x_{i2}, \ldots, x_{ip})$ denote the ith row of the design matrix $\mathbf{X}$, then

$$\hat{y}_i = \hat{\beta}_0 + \hat{\beta}_1 x_{i1} + \hat{\beta}_2 x_{i2} + \cdots + \hat{\beta}_p x_{ip} \quad \text{and} \quad s^2_{\hat{y}_i} = \mathbf{x}'_i (\mathbf{X}'\mathbf{X})^{-1} \mathbf{x}_i \, s^2,$$

with the bounds being

$$\hat{y}_i - t(\alpha/(2n), n - p - 1)\, s_{\hat{y}_i} < E(y_k) < \hat{y}_i + t(\alpha/(2n), n - p - 1)\, s_{\hat{y}_i}.$$

This computation can also be performed using the function `predict`. The difficulty with the resulting bounds is that they are hard to interpret — generally being long tables of points in $(p + 1)$-dimensional space.

One approach to making things a bit more readable is first to identify that p-tuple which is closest to the centroid, $(\bar{x}_1, \bar{x}_2, \ldots, \bar{x}_p)$, of the domain space. Call this point $\mathbf{x}_k = (x_{k1}, x_{k2}, \ldots, x_{kp})$. Once this has been done, generate a suitably sized set of coordinates for which only one variable varies, say X_1. Then, the points about which the intervals are calculated could be $(x_{i1}, x_{k2}, \ldots, x_{kp})$, where $i = 1, 2, \ldots, n$. Although the x_{i1} can be chosen to be the observed values for X_1, for graphing convenience it is preferable to generate an equally spaced set of points that lie within the (observed) domain space.

Here is an illustration; the following code was used to generate Figure 8.8. First, get the fitted values and bounds

```
attach(MultipleReg)
#Calculate distances of each domain triple from the centroid
dist<-sqrt((x1-mean(x1))^2+(x2-mean(x2))^2+(x3-mean(x3))^2)
```

#Find the point closest to the centroid

```
k<-1; while (dist[k]!=min(dist)){k<-k+1}
```

```
xk<-c(x1[k],x2[k],x3[k])
```

#Set the number of intervals and joint significance

```
n<-length(x1); alpha<-.05
```

#Obtain upper and lower limits of x1

```
a<-round(min(x1),2); inc<-(round(max(x1),2)-a)/n
```

```
b<-a+(n-1)*inc
```

#Store coordinates for estimating intervals[4]

```
xm<-data.frame(x1=seq(a,b,inc),
          x2=rep(xk[2],n),x3=rep(xk[3],n))
```

```
detach(MultipleReg)
```

#Obtain fitted values and bounds

```
results<-predict(Multiple.mod,xm,
       interval="confidence",level=1-alpha/n)
```

Then plot the details using the function `matplot`

```
matplot(xm[,1],results,xlab="x1",ylab="E(y)",
          #Specify plot and line types and line colors
          type="l",lty=c(1,3,3),col="black")
```

#Plot the centroid

```
with(MultipleReg,points(xk[1],y[x1==xk[1]],pch=18))
```

If the argument `col="black"` is left out of the `matplot` function call, the plotted curves are assigned different colors. The argument `pch=18` in the `points` function call instructs R on what type of symbol to use for plotting the centroid in the X_1 $E(Y)$-plane.

---

[4] Be sure to use the same variable names used in the original dataset, in this case `x1`, `x2`, and `x3`; otherwise, things go wrong in the `predict` function. To make things worse, there are no warnings!

8.8.4   Prediction regions

The process is analogous to that for confidence regions for the mean response using the function `predict`, with the exception being that the argument `interval="prediction"` is used.

In preparing code for Sheffe intervals, be sure to replace $s_{\hat{y}_i}$ with

$$s^2_{\hat{y}_{new}} = \left(1 + \mathbf{x}'(\mathbf{X}'\mathbf{X})^{-1}\mathbf{x}\right) s^2$$

and make sure to use

$$\sqrt{m\, F\left(\alpha, m, n - p - 1\right)}$$

in place of

$$\sqrt{(p+1)\, F\left(\alpha, p + 1, n - p - 1\right)}$$

in the calculations.

Chapter 9

Additional Diagnostics for Multiple Regression

9.1   Introduction 203
9.2   Detection of Structural Violations 204
     9.2.1   Matrix scatterplots 204
     9.2.2   Partial residual plots 204
     9.2.3   Testing for lack of fit 205
9.3   Diagnosing Multicollinearity 208
     9.3.1   Correlation coefficients 209
     9.3.2   Variance inflation factors 210
9.4   Variable Selection 211
     9.4.1   Contribution significance of individual variables ... 211
     9.4.2   Contribution significance of groups of variables 211
     9.4.3   The drop1 function 213
     9.4.4   The add1 function 213
     9.4.5   Stepwise selection algorithms 214
9.5   Model Selection Criteria 215
     9.5.1   Mean square error and residual sum of squares 218
     9.5.2   Coefficient of determination 218
     9.5.3   Adjusted coefficient of determination 219
     9.5.4   Mallow's statistic 219
     9.5.5   Akaike and Bayesian information criteria 220
     9.5.6   PRESS statistic 221
9.6   For the Curious 222
     9.6.1   More on multicollinearity 222
         9.6.1.1   Background discussion 222
         9.6.1.2   The condition number 224
         9.6.1.3   Condition indices 225
     9.6.2   Variance proportions 226
     9.6.3   Model selection criteria revisited 228

9.1   Introduction

As might be expected, the inclusion of more than one explanatory variable in a linear regression model adds to the complexity of the analyses of such models. The assessment of four issues are addressed in this chapter: structural violations, multicollinearity, variable selection, and model selection.

For the most part, the model for the dataset `MultipleReg` from Chapter 8 is used in the following illustrations. Where appropriate, additional datasets are generated. Begin this chapter by first clearing the work space, then recall the dataset `MultipleReg` using a command of the form

```
source("Z:/Docs/RCompanion/Chapter8/Data/Data08x01.R")
```

At this point, "MultipleReg" should be the only object in the workspace.

9.2   Detection of Structural Violations

The general structure of multiple linear regression models considered assumes that the response, Y, varies linearly (in an approximate sort of way) with each of the explanatory variables, X_j, $j = 1, 2, \ldots, p$. Quite often this may not be the case and the matrix scatterplot seen earlier allows a visual inspection of whether this is the case. This plot can *sometimes* provide information on whether transformations of explanatory variables, or the inclusion of interactive terms such as products of two or more variables or powers of variables might be useful in improving fit. However, there is a problem; it is usually not very easy to see a multivariate trend in a collection of two-dimensional scatterplots.

9.2.1   Matrix scatterplots

Figure 9.1 can be generated using either of the commands shown below.

```
plot(MultipleReg)
with(MultipleReg,pairs(cbind(y,x1,x2,x3)))
```

Preferred plots between the explanatory variables and the response variable in the first row should not demonstrate clear nonlinear trends, and plots for pairs of explanatory variables should not show clear linear trends.

9.2.2   Partial residual plots

Package `faraway` has a function to produce *partial residual plots* for linear models. These plots are helpful in detecting the presence of nonlinearity in the data with respect to the relationship between the response and an explanatory variable. See, for example, [21, pp. 72–74] or [50, pp. 146–150]. Figure 9.2 was generated using the following code.[1]

```
Multiple.mod<-lm(y~x1+x2+x3,MultipleReg)
require(faraway)
for (i in 1:3) {prplot(Multiple.mod,i)}
detach(package:faraway)
```

---

[1] Note that the functions `win.graph` and `par` were used to prepare the window for the three plots in Figure 9.2.

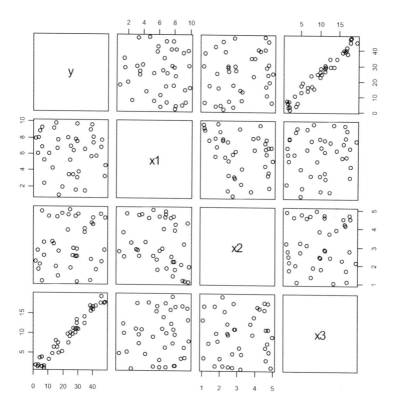

FIGURE 9.1: Matrix scatterplot of data set `MultipleReg` showing trends that may (or may not) exist between (X_j, Y) and (X_j, X_k) pairs.

As one should expect for the data in question, these plots support observations from the matrix scatterplot that there are no noticeable nonlinear trends for (X_j, Y) pairs.

9.2.3 Testing for lack of fit

There are occasions when the data are not appropriate for a linear model.[2] Moreover, because of the number of variables present, simple methods of assessment may prove inadequate for easily identifying undesirable structural features in the data. A test for *lack of* (linear) *fit* (*LOF*) can be performed under certain conditions; see, for example, [45, pp. 119–127].

---

[2] Recall that here the term linear refers to the manner in which the parameters enter the model.

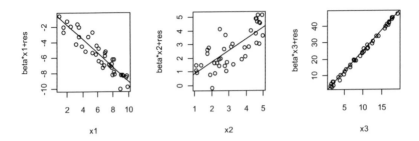

FIGURE 9.2: Partial residual plots for each explanatory variable, obtained using function `prplot` from package `faraway`.

For this test, the model

$$y_i = \beta_0 + \beta_1 x_{i1} + \cdots + \beta_p x_{ip} + \varepsilon_i$$

is assumed as the reduced model (or null hypothesis), and the full model (or alternative hypothesis) is allowed to have some other (very) general form,

$$y_{i,I} = \mu_I + \varepsilon_{i,I},$$

where the index I identifies distinct cells or levels associated with the explanatory variables and i identifies replicates of the observed responses within cells. The computational requirements for testing the above hypotheses are:

1. Every cell I must have at least one observed response; and

2. At least one cell must have more than one (replicate) observed response.

Further assumptions for the LOF test are that the observed responses are normally distributed with (equal) constant variances within and across levels.
    In general, the code to perform an LOF test takes on the form

```
reduced.mod<-lm(y~x1+x2+···+xp, data )
full.mod<-lm(y~factor(x1)+factor(x2)+···+factor(xp), data )
anova(reduced.mod,full.mod)
```

Before providing an illustration of using R to perform this test, a couple of very simple examples may help illustrate the *minimum* experiment design requirements.
    Consider data to be used for a simple regression model. Suppose there are n distinct observed responses and at least one level of the explanatory variable has more than one observed response (replicates). Then the LOF test may be conducted on the model

$$y_{ij} = \mu_j + \varepsilon_{ij}.$$

Next, consider data to be used for a model containing two explanatory variables, for which there are n observed responses. Suppose for at least one pair (x_{j1}, x_{k2}) there are at least two replicates of the response variable. Then the LOF test may be conducted on the model

$$y_{ijk} = \mu_{jk} + \varepsilon_{ijk}.$$

The dataset MultipleReg does not allow for the performance of the LOF test; however, cooking up data is not a problem with R! First, generate some x-values, with room for replicates, and then concoct a set of y-values using a function in which the parameters do not enter linearly. For example,

```
x<-round(runif(50,0,4),1)
#Insert some replicates
x<-c(x,x[rbinom(50,4,.5)+1])
#Generate some "observed" responses
y<-round((9-2*x)^x+rnorm(length(x),0,1),1)+4
#Plot to look at data
plot(y~x)
#Store in a data frame
LOFData<-data.frame(y,x)
#Tidy up some
rm(y,x)
#Save, if so desired
dump("LOFData","Z:/Docs/RCompanion/Chapter9/
                          Data/Data09x01.R")
```

The scatterplot of a sample dataset, LOFData, is shown in Figure 9.3. This scatterplot serves as an adequate alert that the data are unsuited for fitting to a model of the form $y_i = \beta_0 + \beta_1 x_i + \varepsilon_i$ (which, of course, is known because the data were generated using a strange function). In fact, if the data are fitted to a model of the form $y_i = \beta_0 + \beta_1 x_i + \varepsilon_i$ and further diagnostics are performed, plenty of alerts arise. But, the purpose here is to demonstrate the LOF test.

Suppose the various assumptions for the LOF test are satisfied. Then,

```
reduced.mod<-lm(y~x,LOFData)
full.mod<-lm(y~factor(x),LOFData)
anova(reduced.mod,full.mod)
```

produces the following test results.

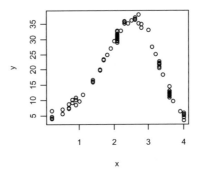

FIGURE 9.3: Scatterplot of data that are probably unsuitable for a linear regression model.

```
Analysis of Variance Table
Model 1: y ~x
Model 2: y ~factor(x)
    Res.Df      RSS   Df  Sum of Sq       F      Pr(>F)
1        98  11253.6
2        70     63.3   28      11190  441.75   < 2.2e-16
```

The row corresponding to model 2 (the alternative) indicates there is plenty of evidence to reject the null hypothesis, so the data are not appropriate for a model of the form $y_i = \beta_0 + \beta_1 x_i + \varepsilon_i$.

Before continuing, remove all objects from the workspace except `MultipleReg` and `Multiple.mod`.

9.3 Diagnosing Multicollinearity

Because of the presence of more than one explanatory variable, linear dependence among explanatory variables becomes a matter of concern. This issue relates to the expectation that the design matrix has full column rank and, while it is very unlikely that the design matrix for a multiple regression model will not have full column rank, there is always a possibility that two or more columns of the design matrix will "almost" be linearly related. This is a consequence of what is referred to as near-collinearity between two or more variables, or more broadly as *multicollinearity*.

While a matrix scatterplot is helpful with respect to pairwise comparisons, it is hard to diagnose linear dependencies among more than two explanatory variables graphically. A variety of measures may be used in diagnosing the presence and strength of "near" linear dependencies among collections of continuous explanatory variables in a multiple regression model; two of these are presented here. Additional approaches and further computational details are presented in the last section of this chapter.

For the data and model in question, execute the following commands:

```
rXX<-with(MultipleReg,cor(cbind(x1,x2,x3)))

rXX.inv<-solve(rXX)

vifs<-numeric(dim(rXX.inv)[1]);vars<-NULL

for (i in 1:3){vifs[i]<-rXX.inv[i,i]
                vars<-c(vars,paste("x",i,sep=""))}

names(vifs)<-vars
```

Here is a breakdown of what has been obtained. The object rXX contains the *correlation coefficient matrix* $\mathbf{r_{X'X}}$ of the variables x_1, x_2, x_3. Next, the R function to invert a matrix is solve; thus, rXX.inv represents the inverse $\mathbf{r_{X'X}^{-1}}$ of the correlation coefficient matrix $\mathbf{r_{X'X}}$. The diagonal entries of $\mathbf{r_{X'X}^{-1}}$ contain what are referred to as the *variance inflation factors*, $(vif)_j$, of the corresponding model. Note that both packages faraway and car contain a function, vif, that can be used to obtain variance inflation factors for a model.

9.3.1 Correlation coefficients

The correlation coefficient matrix of the explanatory variables has the appearance

$$\mathbf{r_{X'X}} = \begin{bmatrix} 1 & r_{12} & \cdots & r_{1p} \\ r_{21} & 1 & \cdots & r_{np} \\ \vdots & \vdots & \ddots & \vdots \\ r_{p1} & r_{p2} & \cdots & 1 \end{bmatrix},$$

where each off-diagonal entry, r_{jk}, is the *correlation coefficient of X_j and X_k* and measures the strength of the linear relationship between the pair X_j and X_k. Values close to 0 are desirable where the matter of near collinearity between pairs of explanatory variables is in question.

A starting point for assessing the presence of multicollinearity is to look at all correlation coefficients r_{ij}, $i \neq j$, to determine if linear dependencies between pairs of explanatory variables might be an issue. The correlation coefficient matrix, rXX, obtained earlier gives

```
> round(rXX,3)
```

```
        x1      x2      x3
x1   1.000  -0.281   0.104
x2  -0.281   1.000   0.058
x3   0.104   0.058   1.000
```

Keep in mind that some nonlinear relationships between explanatory variable pairs can also give correlation coefficient values that are suspiciously close to -1 or 1, so it is best to supplement numerical work with graphical explorations, such as Figure 9.1 for this case. Also, what constitutes a *strong, moderate,* or *weak correlation* often can be discipline-specific. Refer to the literature within your discipline for interpretations of correlation coefficients.

The function cor provides the ability to compute correlation coefficients by three different methods, where the Pearson method used above and described in the last section of this chapter is the default.

9.3.2 Variance inflation factors

Let $k = 1, 2, \ldots, p$, then it can be shown that the diagonal entries of $\mathbf{r}_{\mathbf{X'X}}^{-1}$, called the *variance inflation factors*, have the form

$$vif_k = \frac{1}{1 - R_k^2}$$

where R_k^2 denotes the coefficient of multiple determination for the regression of X_k on the remaining explanatory variables. As suggested (partially) by the name, variance inflation factors provide measures of inflation of the variances of parameter estimates. Observe that a large vif_k is associated with $R_k^2 \approx 1$, when X_k is well explained by a linear combination of the remaining explanatory variables. Conversely, a vif_k close to 1 is associated with $R_k^2 \approx 0$, when X_k is *not* well explained by a linear combination of the remaining explanatory variables. Consequently, yardsticks for variance inflation factors are:

1. If $vif_k \approx 1$, then X_k is not involved in a multicollinearity relationship.

2. If $vif_k > 5$, then X_k is involved in a linear relationship — the variance of $\hat{\beta}_k$ will be large, and β_k is poorly estimated.

For the model in question,

```
> vifs
      x1        x2        x3
1.103201  1.094884  1.019621
```

which all look reasonably close to one.

See the last section of this chapter for further details and methods relating to the multicollinearity issue.

9.4   Variable Selection

Addressing the presence of multicollinearity can be thought of as forming part of the variable selection process in that redundant variables may be identified. Dealing with variables involved in multicollinearity relationships can be sticky, so it is best to refer to discipline-specific literature to find out accepted practices.

Formal measures of variable contribution significance provide another means for selecting appropriate variables.

9.4.1   Contribution significance of individual variables

Consider the object `coefficients` in the model summary object for `Multiple.mod`

```
> Multiple.sum<-summary(Multiple.mod)
> round(Multiple.sum$coefficients,4)
              Estimate   Std. Error   t value   Pr(>|t|)
(Intercept)    2.9411       0.6680     4.4029    1e-04
x1            -0.9098       0.0629   -14.4719    0e+00
x2             0.8543       0.1274     6.7067    0e+00
x3             2.4917       0.0274    90.7870    0e+00
```

The rows corresponding to `Intercept`, `x1`, `x2`, and `x3` provide measures of significance for the corresponding parameter *with the assumption that the remaining parameters are present in the model.*

Recall that the last three rows of the above t-tests are equivalent to testing the contribution significances of each variable *with the assumption that the remaining variables are included in the model.* Because of this, a variable may very well show a low contribution significance but have a significant explanatory effect on the response if tested by itself. Note also that unexpected parameter estimates and test results in the above summary can be an indication of the presence of multicollinearity.

9.4.2   Contribution significance of groups of variables

As already seen, partial F-tests on fitted models can be performed in R by using the function `anova` to compare a reduced model against a proposed full model by means of the *extra sum of squares principle*. Thus, a partial F-test measures the contribution significance of those variables present in a proposed (full) model that are not contained in a given reduced model.

It should be noted that a partial F-test can be used to produce results that are equivalent to the above `coefficients` summary. For example, to test the contribution significance of x2, the code

```
> red.mod<-lm(y~x1+x3,MultipleReg)
> anova(red.mod,Multiple.mod)
Analysis of Variance Table
Model 1: y ~x1 + x3
Model 2: y ~x1 + x2 + x3
  Res.Df    RSS  Df  Sum of Sq       F     Pr(>F)
1     37  72.202
2     36  32.098   1     40.104  44.979  7.988e-08
```

produces an equivalent test for the contribution significance of x2 to the model. Note that the square of the `t.value` for x2 from the `coefficients` summary in `Multiple.sum` equals F (allowing for rounding errors) in the line for model 2 of the partial F-test output. The output under `Pr(>F)` represents the p-value obtained for the test. The next example provides interpretations of the remaining output.

The more practical use of partial F-tests involves testing the (combined) contribution significance of two or more variables to a model. For example, to test the (combined) contribution significance of x1 and x3, execute

```
> red.mod<-lm(y~x2,MultipleReg)
> anova(red.mod,Multiple.mod)
Analysis of Variance Table
Model 1: y ~x2
Model 2: y ~x1 + x2 + x3
  Res.Df    RSS  Df  Sum of Sq       F     Pr(>F)
1     38  7389.3
2     36    32.1   2     7357.2  4125.8  < 2.2e-16
```

The combined contribution significance of the variables x1 *and* x3 is indicated in the row corresponding to model 2: The residual sum of squares (RSS = 7389.3) for model 1 is decomposed into two parts: a sum of squares attributed to the variables x1 *and* x3 (Sum of Sq = 7357.2) and what is left (RSS = 32.1), which is the residual sum of squares for model 2. The F-statistic (F= 4125.8) in the row for model 2 then measures the contribution significance of x1 *and* x3 to the full model.

9.4.3 The drop1 function

Once a model has been constructed, it is also possible to look at the effect of dropping single variables from the model. Rather than using appropriate partial F-tests several times, one may use the R function `drop1` to obtain equivalent information. The basic command for this function is

```
drop1(model,scope,test="F")
```

The argument *model* is the model under consideration and *scope* is a formula containing terms from *model* that are to be tested for contribution significance. If *scope* is not included, all variables are tested. By default, the *Akaike information criterion* (*AIC*) is computed for each model. If `test="F"` is specified, the contribution significance for each dropped variable is also included in the output as an F-test. A model with a smaller AIC (and larger F-statistic) is preferred over a model with a larger AIC (and smaller F-statistic). Consider the model under discussion.

The formula for `Multiple.mod` is `y~x1+x2+x3`, and

```
> drop1(Multiple.mod,~x1+x3,test="F")
Single term deletions
Model:
y ~x1 + x2 + x3
           Df  Sum of Sq     RSS      AIC  F value         Pr(F)
<none>                      32.1   -0.804
x1          1      186.7   218.8   73.977   209.44     < 2.2e-16
x3          1     7348.9  7381.0  214.711  8242.28     < 2.2e-16
```

displays the contribution significances of each of the variables `x1` and `x3` to the full model. Notice that square roots of the F-statistics for each variable match (approximately) with the absolute values of the corresponding t-statistics in the earlier `coefficients` summary in `Multiple.sum`.

9.4.4 The add1 function

One may use the function `add1` to assess the contribution significance of a new variable, proposed to be added to a given model The basic command for this function is similar to the `drop1` function.

```
add1(model,scope,test="F")
```

The arguments *model* and `test="F"` serve the same purpose, but *scope* includes additional variables that are to be tested. The argument *scope* must always be included with at least one new variable.

Suppose the effect of including a (new) fourth variable, say `x4 = x1*x2`, to `Multiple.mod` is of interest. Then,

```
> add1(Multiple.mod,~.+I(x1*x2),test="F")
Single term additions
Model:
y ~x1 + x2 + x3
                Df    Sum of Sq       RSS      AIC  F value  Pr(F)
<none>                         32.098 -0.80362
I(x1 * x2) 1   0.00024849    32.098 1.19607     3e-04   0.987
```

displays the contribution significance of x4=I(x1*x2) to the model
y~x1+x2+x3+I(x1*x2).

It can be observed that

```
add1(Multiple.mod,~x1+x2+x3+I(x1*x2),test="F")
```

produces identical results. The period "." in the previous add1 call instructs
R to use all variables already present in the current model.

9.4.5   Stepwise selection algorithms

Another useful R function, particularly when dealing with large numbers of
explanatory variables, is the function step which selects the best model using
a chosen stepwise selection algorithm by means of the AIC selection criterion.
The simplest call for this function contains the proposed model formula as its
only argument. More details can be extracted by including some additional
arguments, for example, in

```
step(model,direction,trace)
```

the argument direction may be specified as direction="both", "backward",
or "forward"; the default setting is "both". Assigning a nonzero (positive)
value to trace, such as trace=1, 2 or higher, will result in information output
during the running of the function. A value of 1 is assigned by default, and
a value of 0 instructs R to output parameter estimates associated with only
those variables that are eventually selected for inclusion in the model. Thus,

```
> step(Multiple.mod)
Start: AIC=-0.8
y ~ x1 + x2 + x3
            Df  Sum of Sq      RSS      AIC
<none>                  32.1   -0.804
- x2    1          40.1    72.2   29.623
- x1    1         186.7   218.8   73.977
```

```
 - x3     1       7348.9   7381.0   214.711
Call:
lm(formula = y ~x1 + x2 + x3, data = MultipleReg)
Coefficients:
(Intercept)         x1       x2       x3
     2.9411    -0.9098   0.8543   2.4917
```

produces the results of a stepwise selection, the output appearing to be that of the backward selection process. Be aware that the output will vary, depending on the arguments passed into the function and the number of variables proposed for the model. Placing `test="F"` in the function call results in the contribution significance for each dropped variable being included in the output as an F-test.

The row corresponding to `<none>` provides measures for the full model, the row corresponding to `- x2` provides measures for the model not including x2, and so on. The model having the lowest `AIC` is selected and coefficient estimates for the chosen model are given in the last row of the output.

9.5   Model Selection Criteria

A part of the model selection process does involve variable selection; however, there are times when some of the above variable selection processes result in a pool of two or more candidate models. Some fundamental principles/goals associated with the construction of a multiple linear regression model are as follows:

1. Make use of the underlying system (physical, biological, etc.) from which the data are obtained. This might involve features such as the signs or magnitudes of a proposed model's parameters.

2. Explore structural features suggested by the data, identify the more important explanatory variables, and attempt to distinguish, in particular, between those involved in multicollinearity.

3. Strive for simplicity, seek a model with the fewest explanatory variables without sacrificing explanatory power. Models that are additive in the variables included are preferred over those in which products or powers of variables are present.

4. Identify the simplest model that provides the best predictive power.

The focus of this section is Item 4. Some of the criteria to be used have already been encountered.

Given an appropriate dataset, any model (of the form under discussion) to which the data are fitted uses variables from a pool of p explanatory variables $\{X_1, X_2, \ldots, X_p\}$. For each such model, a collection of selection criteria can be computed and used to determine which model might be the best for the purpose in mind. Although the typical number of candidate models is small, here an exhaustive comparison is performed with the current dataset for purposes of illustration.

The last section of this chapter presents an example of a program that generates model selection criteria. However, the simplest strategy at this point is to make use of a nifty function, regsubsets, contained in package leaps [47]. Using the current dataset, begin with

```
require(leaps)
#Use the function regsubsets to get details on all possible models
search.results<-regsubsets(y~x1+x2+x3,data=MultipleReg,
                    method="exhaustive",nbest=3)
#Extract search criteria from the object search.results
selection.criteria<-summary(search.results)
detach(package:leaps)
```

The argument method="exhaustive" instructs the function regsubsets to do an exhaustive search of all possible models, and nbest=3 says to pick the three best models from *each of* the 1-, 2-, and 3-variable models.

Of the two objects created above, selection.criteria is the one of interest. This object contains most of the selection criteria to be discussed, and enough information to calculate those that are not included.

```
> names(selection.criteria)
[1] "which" "rsq" "rss" "adjr2" "cp" "bic" "outmat" "obj"
```

The contents of interest are: rsq and adjr2, the coefficient of determination and adjusted coefficient of determination, respectively, for each model; rss, the residual sum of squares; and cp, Mallow's statistics for each model. Information on which variables are contained in each model is stored in which.

Next,

```
#Obtain the sample size
n<-length(MultipleReg[,1])
#Extract the number of variables included in the test model
q<-as.integer(row.names(selection.criteria$which))
#Extract the mean square error for the test model
```

```
mse<-selection.criteria$rss/(n-q-1)
#Extract R square and adjusted R square
R.sq<-selection.criteria$rsq
AdjR.sq<-selection.criteria$adjr2
#Extract Mallow's statistic
Cp<-selection.criteria$cp
#Compute the AIC for the test model by formula
aic.f<-n*log(selection.criteria$rss)-n*log(n)+2*(q+1)
#Compute the BIC for the test model³ by formula
bic.f<-n*log(selection.criteria$rss)-n*log(n)+(q+1)*log(n)
#Extract variable information
var<-as.matrix(selection.criteria$which[,2:4])
#Create the criteria table
criteria.table<-data.frame(cbind(q,mse,R.sq,AdjR.sq,
    Cp,aic.f,bic.f,var[,1],var[,2],var[,3]),row.names=NULL)
#Name the columns
names(criteria.table)<-c("q","MSE","Rsq","aRsq","Cp",
                        "AIC","BIC","x1","x2","x3")
#Clean up the workspace
rm(n,q,mse,R.sq,AdjR.sq,Cp,aic.f,bic.f,var)
#Take a look at the contents of the criteria table
round(criteria.table,2)
```

The results are

	q	MSE	Rsq	ARsq	Cp	AIC	BIC	x1	x2	x3
1	1	8.79	0.96	0.96	338.83	88.91	92.29	0	0	1
2	1	194.45	0.03	0.01	8251.57	212.76	216.13	0	1	0
3	1	199.59	0.01	-0.02	8470.41	213.80	217.18	1	0	0
4	2	1.95	0.99	0.99	46.98	29.62	34.69	1	0	1
5	2	5.91	0.97	0.97	211.44	73.98	79.04	0	1	1
6	2	199.49	0.03	-0.02	8244.28	214.71	219.78	1	1	0
7	3	0.89	1.00	1.00	4.00	-0.80	5.95	1	1	1

---

[3] At the time of writing this Companion, BIC results from `regsubsets` differed from values obtained using the `step` or `extractAIC` functions with k=log(n) as well as if the formulae for computing AIC and BIC are used.

Here are brief discussions on the above measures. For brevity, in what follows a particular model will be identified by the variable indices set. Thus, Model-123 represents the model containing the variables x_1, x_2, and x_3. In the following, let q be the number of variables in the model specified by the index set J (for example, $q = 3$ for $J = 123$). Alternative commands to calculate the selection criteria measures are given, and the above results will be referred to for each case.

9.5.1   Mean square error and residual sum of squares

Let q be the number of variables contained in Model-J, then the *mean square error* (MSE) for each model, denoted by

$$s_J^2 = \frac{SSE_J}{n - q - 1},$$

can be obtained from the model summary object

```
MSEJ<-modelJ.sum$sigma^2
```

and

```
SSEJ<-MSEJ*(n-q-1)
```

gives the *residual sum of squares* (SSE). Recall that `sigma` is the *residual standard error*, s_J, for the model in question. Minimizing these quantities is the goal in selecting the best model. The MSE can be considered a standardized version of the residual sum of squares in the sense that it is adjusted for the number of variables contained in the model under review. From the above results, Model-123 minimizes MSE (and SSE).

9.5.2   Coefficient of determination

The *coefficient of multiple determination*,

$$R_J^2 = 1 - \frac{SSE_J}{SSTo},$$

can be obtained, again, from the model summary object

```
RsqJ<-modelJ.sum$r.squared
```

This serves as *measure of explained variation*. Typically, low values (close to zero) are undesirable and high values (close to one) are desireable, but should be interpreted with some caution. For the dataset in question, Model-123 has the largest R^2 of the seven models.

9.5.3 Adjusted coefficient of determination

The *adjusted coefficient of multiple determination*,

$$R^2_{J\ \text{Adj}} = 1 - \frac{MSE_J}{MST_o},$$

is contained in the model summary object

```
AdjRsqJ<-modelJ.sum$adj.r.squared
```

This also serves as *measure of explained variation*, adjusted for the number of variables. Again, low values (close to zero) are undesirable and high values (close to one) are desirable, but results should be interpreted with some caution. Once again, Model-123 has the largest R^2_{Adj} of the seven models.

9.5.4 Mallow's statistic

Mallow's C_p statistic provides a means for assessing the predictive quality of a model, and is estimated by

$$C_J = \frac{SSE_J}{s^2} + 2(q+1) - n.$$

This can be computed using

```
CJ<-SSEJ/MSE+2*(q+1)-n
```

where `MSE` is the mean square error for the complete model (with all variables included).

This statistic distinguishes between what are referred to as *underspecified models*, those that may have too few variables, and *overspecified models*, those that have variables that are marginal or involved in multicollinearity. An underspecified model suffers from *biased regression coefficient estimates* and *biased prediction* or response estimates. An overspecified model can result in large variances for both coefficient estimates and prediction or fitted responses.

The structure of Mallow's statistic includes a variance and a bias component allowing the following interpretations. For any given model:

1. The closer the value of Mallow's statistic is to the number of parameters, the less the sampling error is.

2. If Mallow's statistic is larger than the number of parameters, then the model is interpreted as having a large bias.

3. If Mallow's statistic is equal to the number of parameters, then the model is interpreted as having no bias. Note that the complete model satisfies this condition even though it may not be the "best" model.

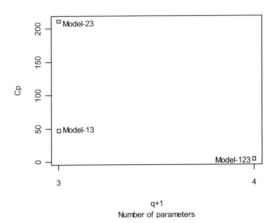

FIGURE 9.4: Plot of lowest three Mallow's statistics against the number of model parameters.

4. If Mallow's statistic is smaller than the number of parameters, then the model is interpreted as having no bias.

In summary, in a plot of C_J (y-axis) against $q + 1$ (x-axis) for all candidate models, the model with $(C_J, q + 1)$ lying closest to (usually below) the line $y = x$ and with the lowest Mallow's statistic is preferred. Typically, very large Mallow's statistics are left out of plots of Mallow's statistics to aid in distinguishing differences between values.

Figure 9.4, which was obtained using the function `stripchart` along with the function `text` for point labels, contains the lowest three Mallow's statistics for the dataset in question. Including a reference line $y = x$ is helpful when there are many good candidates. In this case, however, there is a clear winner; Model-123 wins out using the C_p-criterion.

9.5.5 Akaike and Bayesian information criteria

The *Akaike information criterion* (*AIC*) is given by

$$AIC_J = n \ln(SSE_J) - n \ln(n) + 2(q + 1),$$

and can be obtained using

```
AICJ<-n*log(SSEJ)-n*log(n)+2*(q+1)
```

The *Bayesian information criterion* (*BIC*), also called the *Schwarz–Baysian criterion* (*SBC*), is given by

$$BIC_J = n \ln(SSE_J) - n \ln(n) + (q + 1) \ln(n),$$

and can be obtained using

```
BICJ<-n*log(SSEJ)-n*log(n)+(q+1)*log(n)
```

Both BIC and AIC are minimized in the search for the preferred model. Both criteria favor small SSE_J and penalize models with more variables. However, for larger sample sizes ($n \geq 8$), the BIC imposes a larger penalty on candidate models. Once again, Model-123 wins out under both criteria.

R has a built-in function, `extractAIC`, which is contained in package `stats`, for extracting both BIC and AIC for a given model.

9.5.6 PRESS statistic

The *prediction sum of squares* (*PRESS*) criterion is obtained from the *PRESS* or *deleted residuals* of a model. If $\hat{d}_i$ is the ith deleted residual for a particular model, then

$$PRESS = \sum_{i=1}^{n} \hat{d}_i^2,$$

where the deleted residuals can be computed using

$$\hat{d}_i = y_i - \hat{y}_{(i)} = \frac{\hat{\varepsilon}_i}{1 - h_{ii}}.$$

The studentized version of these residuals was seen earlier in identifying (influential) outlying observations. Like studentized deleted residuals, individual PRESS residuals provide measures of the stability of a particular fitted regression model and may be used to identify observed responses that influence estimated responses. PRESS residuals with relatively large magnitudes signal points of high influence, which may be potentially damaging to the predictive power of a model. So, the PRESS statistic for a model serves as an overall measure of the predictive "quality" of a model and a lower PRESS statistic is desirable.

Consider looking at the PRESS statistics for Model-13 (contains x1 and x3), Model-23 (contains x2 and x3), and Model-123 (the complete model). A basic set of code can be put together

```
mods<-c("y~x1+x3","y~x2+x3","y~x1+x2+x3")
P<-numeric(3)
for (i in 1:3){mod<-lm(mods[i],MultipleReg)
               r<-residuals(mod)
               h<-hatvalues(mod)
               d<-r/(1-h)
               P[i]<-sum(d^2)}
P<-data.frame("PRESS"=P,row.names=mods); P
```

and the results are

$$PRESS$$

	PRESS
y~x1+x3	84.97148
y~x2+x3	251.37813
y~x1+x2+x3	39.22409

As might be expected, the results suggest Model-123 is the "best," which agrees with all previous criteria. This (agreement of all criteria) is not necessarily always going to be the case.

Package MPV [8] contains a function PRESS that computes the PRESS statistic of a given model. Using this function in the above code removes the need to compute r, h, d, and sum(d^2).

9.6   For the Curious

The first part of this section delves deeper into investigations of the multicollinearity issue. In the second part, alternative code for extracting and compiling model selection criteria is presented.

9.6.1   More on multicollinearity

Understanding the mechanics of what follows probably requires a fair understanding of linear algebra; however, here is an attempt at an abbreviated and somewhat informal view of things. Further reading on the subject matter can be found in, for example, [5, Ch. 3], [53, Ch. 7], and [45, Sec. 7.5–7.6], and a fairly rigorous treatment of principal components is given in [42, Ch. 8]. The notation and discussion in this section follow along the lines of a combination of the above works.

First set the stage.

9.6.1.1   Background discussion

For $j = 1, 2, \ldots, p$ and $i = 1, 2, \ldots, n$ let Y, X_j, y_i, and x_{ij} be as already defined. Denote the sample means of Y and X_j by $\bar{y}$ and $\bar{x}_j$, respectively, and their respective standard deviations by s_y and s_{x_j}. Begin by transforming the data using what is referred to in [45] as the *correlation transformation*.

For $j = 1, 2, \ldots, p$ and $i = 1, 2, \ldots, n$, compute

$$y_i^* = \frac{y_i - \bar{y}}{s_y \sqrt{n-1}} \quad \text{and} \quad x_{ij}^* = \frac{x_{ij} - \bar{x}_j}{s_{x_j} \sqrt{n-1}}$$

to obtain the transformed data, then fit this transformed data to the (*centered* and *scaled*) model

$$y_i^* = \alpha_1 x_{i1}^* + \alpha_2 x_{i2}^* + \cdots + \alpha_p x_{ip}^* + \varepsilon_i^*.$$

In matrix form, this model has the appearance

$$\mathbf{y}^* = \mathbf{X}^* \boldsymbol{\alpha} + \boldsymbol{\varepsilon}^*$$

and the corresponding *normal equation* is

$$\mathbf{X}^{*\prime} \mathbf{X}^* \hat{\boldsymbol{\alpha}} = \mathbf{X}^{*\prime} \mathbf{y}^*.$$

Parameter estimates are then obtained, as with the untransformed model, by left-multiplying both sides of the normal equation by the inverse of $\mathbf{X}^{*\prime}\mathbf{X}^*$,

$$\hat{\boldsymbol{\alpha}} = \left(\mathbf{X}^{*\prime} \mathbf{X}^* \right)^{-1} \mathbf{X}^{*\prime} \mathbf{y}^*.$$

With some effort it can be shown that $\hat{\boldsymbol{\alpha}}$ for the transformed model and $\hat{\boldsymbol{\beta}}$ for the untransformed model are (linearly related) scaled versions of each other. The key point here is that the matrix $\mathbf{X}^{*\prime}\mathbf{X}^*$ and the correlation coefficient matrix $\mathbf{r}_{\mathbf{X}'\mathbf{X}}$ of the variables $X_1, X_2, \ldots, X_p$ are one and the same. From here on, only the notation $\mathbf{r}_{\mathbf{X}'\mathbf{X}}$ will be used for consistency.

In order to compute $\hat{\boldsymbol{\alpha}}$, the matrix $\mathbf{r}_{\mathbf{X}'\mathbf{X}}$ needs to be invertible. So, as long as $\mathbf{r}_{\mathbf{X}'\mathbf{X}}$ does not posses certain other "bad characteristics" (as long as it is *well-conditioned*), entries in its inverse are not subject to "blowups" or severe roundoff errors. This ensures stable parameter estimates. So, a well-conditioned $\mathbf{r}_{\mathbf{X}'\mathbf{X}}$ produces a "nice" $\mathbf{r}_{\mathbf{X}'\mathbf{X}}^{-1}$ containing variance inflation factors that most likely will lie within acceptable bounds. This ensures stable estimates for $\boldsymbol{\alpha}$ (and, hence, $\boldsymbol{\beta}$). With this in mind, it would be nice to have some way to assess the "conditioning" of a correlation matrix. For this, some further terminology and notation are needed.

Let $\mathbf{v}$ be a nonzero p-dimensional vector and λ a real number such that the relation

$$\mathbf{r}_{\mathbf{X}'\mathbf{X}} \, \mathbf{v} = \lambda \, \mathbf{v}$$

holds true. Then λ is called an *eigenvalue* of $\mathbf{r}_{\mathbf{X}'\mathbf{X}}$ and $\mathbf{v}$ is called the *eigenvector* (of $\mathbf{r}_{\mathbf{X}'\mathbf{X}}$) corresponding to λ. The pair $(\lambda, \mathbf{v})$ is referred to as an *eigenpair* of $\mathbf{r}_{\mathbf{X}'\mathbf{X}}$ and an analysis of the eigenpairs of $\mathbf{r}_{\mathbf{X}'\mathbf{X}}$ constitutes what is referred to as an *eigenanalysis* of $\mathbf{r}_{\mathbf{X}'\mathbf{X}}$.

Because of certain properties that (any) correlation coefficient matrix possesses, there are certain guarantees. The matrix $\mathbf{r}_{\mathbf{X}'\mathbf{X}}$ is guaranteed to have p real-valued non-negative eigenvalues, which are not necessarily distinct. Moreover, in the case where two or more explanatory variables have an exact linear relationship,[4] at least one of the eigenvalues will be zero. If an eigenvalue of

---

[4] In R, such variables are classified as "aliased" (dependent) variables. The matter of aliases plays a big role in linear models with categorical variables, to be discussed later.

$\mathbf{r_{X'X}}$ is zero, the inverse of $\mathbf{r_{X'X}}$ does not exist ($\mathbf{r_{X'X}}$ is said to be *singular*) and the discussion ends. Note that a zero eigenvalue is rarely encountered in regression models and would typically be a (very rare) consequence of oversight in the planning phase of a study.

As long as there are no *exact* linear relationships between two or more explanatory variables, none of the eigenvalues equal zero and the matrix $\mathbf{r_{X'X}}$ is invertible. With respect to the conditioning of $\mathbf{r_{X'X}}$, it turns out that the eigenvalues of $\mathbf{r_{X'X}}$ provide a means for measuring this characteristic. Suppose a given $\mathbf{r_{X'X}}$ has

$$\lambda_1 \geq \lambda_2 \geq \cdots \geq \lambda_p > 0$$

as its eigenvalues; R stores and outputs eigenvalues in descending order. Also, be aware that the subscripts for the eigenvalues have nothing to do with variable subscripts. So, for example, λ_1 is not an eigenvalue associated with x_1, and so on. In computing the inverse, $\mathbf{r_{X'X}^{-1}}$, the reciprocals of the above eigenvalues appear within $\mathbf{r_{X'X}^{-1}}$. So, (very) small eigenvalues will tend to create blowups or rounding errors in the entries of $\mathbf{r_{X'X}^{-1}}$; these in turn create instabilities in the estimates for $\boldsymbol{\alpha}$ (and, hence, $\boldsymbol{\beta}$). This idea motivates the first measure in this discussion.

9.6.1.2　The condition number

The *condition number* of $\mathbf{r_{X'X}}$, defined by

$$\phi = \lambda_{\max} / \lambda_{\min},$$

is used to measure, in a sense, the "size" or "magnitude" of $\mathbf{r_{X'X}}$. Notice that $\phi \geq 1$ and, ideally, we would like ϕ to be close to 1 in value to ensure that the smallest eigenvalue is not considerably smaller than the largest. Thus, a desireable scenario is one in which all the eigenvalues are close in value, and a large range of values is an indication of the possible presence of multicollinearity.

Using rXX from earlier, eigenvalues for the correlation coefficient matrix can be obtained using

```
> (eigenvals<-eigen(rXX)$values)
[1] 1.2849666 1.0372412 0.6777922
```

Notice that the eigenvalues are fairly close in value. Also, curiously,

```
> sum(eigenvals)
[1] 3
```

an interesting property of the eigenvalues for matrices having certain properties that $\mathbf{r_{X'X}}$ happens to have. It turns out that the sum of the eigenvalues of such matrices always equals the number of columns in the matrix. Now, the condition number of $\mathbf{r_{X'X}}$ is

```
> (c.number<-max(eigenvals)/min(eigenvals))
[1] 1.895812
```

and the question is whether this lies within acceptable bounds. Yardsticks[5] for ϕ are:

1. If $\phi \approx 1$ or close to 1, multicollinearity is not a problem.

2. If $100 < \phi < 1000$, then moderate multicollinearity may be present.

3. If $\phi > 1000$, then severe multicollinearity is very likely to be present.

A large condition number is an indication that the inverse of $\mathbf{r_{X'X}}$ involves divisions by very small numbers, leading to instabilities in parameter estimates. Near linear dependencies between two or more variables can then result in parameter estimates being highly sensitive to small changes in the data.

9.6.1.3 Condition indices

Condition indices[6] are measures that focus on specific eigenvectors (or *principal components*) of $\mathbf{r_{X'X}}$, and are defined by[7]

$$\phi_j = \lambda_{\max}/\lambda_j.$$

For the model and data in question,

```
> (c.indices<-max(eigenvals)/eigenvals)
[1] 1.000000 1.238831 1.895812
```

Typically, if any multicollinearity is present it is indicated in the right tail of this list.

Yardsticks for condition indices follow from the condition number, all eigenvectors (also referred to as *principal components*) having a condition index greater than 100 should be suspected as possesing multicollinearity. The definition of eigenpairs permits a way to squeeze some further information out of knowing a particular eigenvector possesses multicollinearity.

Suppose one of the condition indices is large, say $\phi_k > 1000$, and let $\mathbf{v}_k' = (v_{1k}, v_{2k}, \cdots, v_{pk})$ be the eigenvector (principal component) corresponding to $\lambda_k \approx 0$. Then, by definition,

$$\mathbf{r_{X'X}}\,\mathbf{v}_k = \lambda_k\mathbf{v}_k \approx \mathbf{0},$$

---

[5] Some define the condition number as the square root of $\lambda_{\max}/\lambda_{\min}$. In this scenario, the square roots of the given yardsticks apply.

[6] These are also referred to as condition numbers by some.

[7] As with the condition number, some define condition indices to be the square roots of $\lambda_{\max}/\lambda_j$. Again, the square roots of the yardsticks given for the condition number apply.

and it can be shown that this leads to

$$v_{1k}X_1^* + v_{2k}X_2^* + \cdots + v_{pk}X_p^* \approx 0,$$

where $X_1^*, X_2^*, \cdots, X_p^*$ represent standardized (scaled and centered) versions of the original explanatory variables. By removing those variables, X_j^*, having small coefficients, v_{jk}, this (approximate) linear relation can sometimes prove useful in identifying (scaled and centered) variables that are "nearly" linearly related.

Eigenvectors for $\mathbf{r_{X'X}}$ can be obtained as follows:

```
> (eigenvecs<-eigen(rXX)$vectors)
            [,1]          [,2]         [,3]
[1,]   0.7192428   0.1502101  0.6783264
[2,]  -0.6836364   0.3270170  0.6524577
[3,]   0.1238185   0.9330042 -0.3378937
```

It is useful to remember that the eigenvector, $\mathbf{v}_k$, corresponding to the eigenvalue λ_k is stored in the kth column of the object `eigenvecs` created above. Also, keep in mind that caution should be exercised when choosing which variables (if any) to remove from a model.

9.6.2   Variance proportions

The last useful piece of the multicollinearity issue, at this level, addresses determining what proportion of the variance of each parameter estimate is attributed to each (potentially severe) linear dependency. For this analysis, the data should be scaled but *should not be centered*.

To set the stage, denote the first column (of 1s) in the design matrix by X_0; then, for $i = 1, 2, \ldots, n$ and $j = 0, 1, \ldots, p$, construct the scaled design matrix, $\mathbf{Z}$, having entries

$$z_{ij} = x_{ij} \Big/ \sqrt{\sum_{i=1}^{n} x_{ij}^2}$$

and compute $\mathbf{Z'Z}$. An eigenanalysis of $\mathbf{Z'Z}$ then provides a view of things *with the intercept term included*. While much of what was mentioned above applies, here things are taken a step further.

First, obtain $\mathbf{Z'Z}$ and then the eigenvalues, eigenvectors, and condition indices of $\mathbf{Z'Z}$.

```
attach(MultipleReg)
x0<-rep(1,40)
Z<-cbind(x0/sqrt(sum(x0^2)),x1/sqrt(sum(x1^2)),
```

```
                    x2/sqrt(sum(x2^2)),x3/sqrt(sum(x3^2)))
detach(MultipleReg)
ZTZ<-crossprod(Z,Z)
l<-eigen(ZTZ)$values; v<-eigen(ZTZ)$vectors; ci<-max(l)/l
```

See what `sum(l)` produces.

Keeping in mind that there are $p+1$ parameters in the noncentered case, let $(\lambda_0, \mathbf{v}_0), (\lambda_1, \mathbf{v}_1), \ldots, (\lambda_p, \mathbf{v}_p)$ denote the eigenpairs of the matrix $\mathbf{Z'Z}$, where for $j = 0, 1, 2, \ldots, p$ the eigenvectors have the form $\mathbf{v}_j' = (v_{0j}, v_{1j}, \ldots, v_{pj})$. To compute the proportion, p_{jk}, of the variance $s_{\beta_k}^2$ which is attributed to the eigenvalue λ_j, first compute

$$a_k = \sum_{j=0}^{p} \frac{v_{kj}^2}{\lambda_j}, \quad \text{for each } k = 0, 1, 2, \ldots, p,$$

then compute

$$p_{jk} = \frac{v_{kj}^2 \big/ \lambda_j}{a_k},$$

for $k = 0, 1, \ldots, p$ and $j = 0, 1, \ldots, p$. Here is some code with comments:

```
#Initialize objects for a_k and p_jk
a<-numeric(4); p<-cbind(a,a,a,a)
colnames(p)<-c("b0","b1","b2","b3")
#Generate the variance proportions
for (k in 1:4){
        a[k]<-sum(v[k,]^2/l)
        for (j in 1:4){
                p[j,k]<-(v[k,j]^2/l[j])/a[k]}}
#Store the results in a data frame
collinearity<-data.frame("eigenval."=l,"Cond. Index"=ci,p)
```

Now take a look:

```
> round(collinearity,3)
    eigenval. Cond.Index    b0     b1     b2     b3
1     3.603       1.000  0.004  0.009  0.008  0.016
2     0.188      19.182  0.013  0.004  0.165  0.865
3     0.174      20.658  0.001  0.439  0.230  0.092
4     0.035     104.398  0.982  0.548  0.597  0.028
```

Observe that columns under each parameter estimate (`bj`) all sum to 1 (allowing for rounding).

As an illustration of interpreting the variance proportions, consider the column `b3` which corresponds to $\hat{\beta}_3$. The first eigenvalue contributes approximately 2% of the variance of $\hat{\beta}_3$, the second eigenvalue contributes approximately 87%, the third contributes 9%, and the fourth contributes 3%.

Variance proportions provide a means of determining how "damaging" a particular dependency (eigenvalue) is to each parameter estimate. So, a small eigenvalue, which indicates the presence of moderate to serious multicollinearity, deposits its influence, to some degree, on all parameter estimate variances. In cases where a small eigenvalue is accompanied by a collection of two or more coefficients having high variance proportions (50% or more, as suggested in [5]), one may conclude that multicollinearity exists between the corresponding regression variables. The linear dependency between these variables, then, adversely affects the precision of the corresponding parameter estimates.

9.6.3 Model selection criteria revisited

As should be evident by this time, R can be coaxed into performing a fair amount of complex computations. Here, consider preparing a program that performs a task equivalent to that performed by the function `regsubsets`. The program presented illustrates combinations of a variety of features introduced in earlier chapters of this Companion.

First, generate an exhaustive list of model formulas. There are a total of seven models containing at least one variable.

```
#Prepare "switches" for each variable, 1 says include, 0 says no
x1<-c(1,0,0,1,1,0,1)
x2<-c(0,1,0,1,0,1,1)
x3<-c(0,0,1,0,1,1,1)
#Initialize the model formula storage object
mods<-character(7)
#Use a for-loop to generate each model formula
for (i in 1:7){
    mods[i]<-paste("y~",
      ifelse(x1[i]==1,"x1",""),
        ifelse((x1[i]!=0)&((x2[i]!=0)|(x3[i]!=0)),"+",""),
          ifelse(x2[i]==1,"x2",""),
            ifelse((x2[i]==1)&(x3[i]!=0),"+",""),
              ifelse(x3[i]==1,"x3",""),sep="")}
```

Execute mods to make sure the contents are what are wanted. Next, initialize the selection criteria storage objects.

```
#Variable counter
q<-integer(7)
#Residual sum of squares and mean square error
SSE<-numeric(7); MSE<-numeric(7)
#R square and adjusted R square
Rsq<-numeric(7); A.Rsq<-numeric(7)
#Mallow's statistic and PRESS statistic
Cp<-numeric(7); P<-numeric(7)
#AIC and BIC
aic<-numeric(7); bic<-numeric(7)
```

Now generate the various criteria for each model.

```
#Get sample size and mean square error for full model
n<-dim(MultipleReg)[1]; s.sq<-Multiple.sum$sigma^2
#Begin for-loop
for (i in 1:7){
        #Fit model i and get summary
        mod<-lm(mods[i],MultipleReg)
        mod.sum<-summary(mod)
        #Get number of variables
        q[i]<-as.integer(x1[i]+x2[i]+x3[i])
        #Get mean square error and residual sum of squares
        MSE[i]<-round(mod.sum$sigma^2,2)
        SSE[i]<-mod.sum$sigma^2*(n-q[i]-1)
        #Get R square and adjusted R square
        Rsq[i]<-round(mod.sum$r.squared,3)
        A.Rsq[i]<-round(mod.sum$adj.r.squared,3)
        #Get Mallow's statistic
        Cp[i]<-round(SSE[i]/s.sq+2*(q[i]+1)-n)
        #Get AIC and BIC
        aic[i]<-round(n*log(SSE[i])-n*log(n)+2*(q[i]+1))
        bic[i]<-round(n*log(SSE[i])-n*log(n)+log(n)*(q[i]+1))
```

```
#Get PRESS
r<-residuals(mod); h<-hatvalues(mod)
P[i]<-round(sum((r/(1-h))^2))}
```

Finally, store the information:

```
#Create a data frame
Model.selection<-data.frame(
    #Put in q, MSE, R-squared, and adjusted R-squared
    "q"=q, "MSE"=MSE, "R.Sq"=Rsq, "A.Rsq"=A.Rsq,
        #Put in Cp, AIC, BIC, and PRESS
        "Cp"=Cp, "AIC"=aic, "BIC"=bic, "PRESS"=P,
            #Assign row names
            row.names=mods)
```

and take a look

```
> Model.selection
```

	q	MSE	R.Sq	A.Rsq	Cp	AIC	BIC	PRESS
y~x1	1	199.59	0.007	−0.019	8470	214	217	8311
y~x2	1	194.45	0.032	0.007	8252	213	216	8201
y~x3	1	8.79	0.956	0.955	339	89	92	370
y~x1+x2	2	199.49	0.033	−0.019	8244	215	220	8524
y~x1+x3	2	1.95	0.991	0.990	47	30	35	85
y~x2+x3	2	5.91	0.971	0.970	211	74	79	251
y~x1+x2+x3	3	0.89	0.996	0.995	4	−1	6	39

Taking into account that the above measures have been rounded for page-fitting reasons, the values obtained duplicate those obtained earlier.

Chapter 10

Simple Remedies for Multiple Regression

10.1 Introduction 231
10.2 Improving Fit 232
    10.2.1 Power transformations of explanatory variables 232
    10.2.2 Transforming response and explanatory variables 235
10.3 Normalizing Transformations 238
10.4 Variance Stabilizing Transformations 239
10.5 Polynomial Regression 239
10.6 Adding New Explanatory Variables 241
10.7 What if None of the Simple Remedies Help? 242
    10.7.1 Variance issues 242
    10.7.2 Outlier issues 242
    10.7.3 Multicollinearity issues 242
    10.7.4 Autocorrelation issues 243
    10.7.5 Discrete response matters 243
    10.7.6 Distributional issues in general 243
10.8 For the Curious: Box–Tidwell Revisited 243

10.1 Introduction

All concerns and simple remedies applicable to simple regression models apply here: Transformations of the response or explanatory variables to improve fit; transformations of the response variable to address violations of the normality assumption; and transformations of the response variable to stabilize variances. For the most part, the methods and functions described for simple regression models work just fine here, but the presence of two or more explanatory variables can make identifying a suitable transformation of the explanatory variables tricky.

This chapter recalls or defers to methods used for simple regression and brings up a couple of new ideas to overcome difficulties arising from a multidimensional domain space. Finally, some leads are provided on where to go if none of the simple remedies seen in this Companion work.

10.2   Improving Fit

Determining appropriate transformations for the explanatory variables in the case of multiple regression models can be tricky. Underlying theoretical principles of the discipline associated with a model can be helpful and often provide quite a bit of guidance in choosing appropriate transformations. When this is not an option, an exploratory graphical analysis of the data might help provide a partial (though sometimes misleading) picture of what might help. When flying blind, there is a numerical option.

10.2.1   Power transformations of explanatory variables

Let Y and X_j, $j = 1, 2, \ldots, p$ be as defined for the multiple regression model introduced in Chapter 8. The procedure in question, proposed by Box and Tidwell and called the Box–Tidwell procedure, obtains estimates for exponents $\lambda_1, \lambda_2, \ldots, \lambda_p$ to result in a transformed model of the form

$$y_i = \beta_0 + \beta_1 w_{i1} + \cdots + \beta_p w_{ip} + \varepsilon_i$$

where

$$w_{ij} = \begin{cases} x_{ij}^{\lambda_j} & \text{if } \lambda_j \neq 0, \\ \ln(x_{ij}) & \text{if } \lambda_j = 0, \end{cases}$$

and for which the fit is an improvement over a preliminary, but potentially poorly fitting model of the form

$$y_i = \beta_0 + \beta_1 x_{i1} + \cdots + \beta_p x_{ip} + \varepsilon_i.$$

Very briefly, the procedure involves applying a Taylor Series expansion of Y about $(\lambda_1, \lambda_2, \ldots, \lambda_p)$ iteratively until convergence, up to a desired tolerance level, is reached. Details on the procedure and a set of step-by-step instructions are given in [53, pp. 192–194]. Sample code to implement this procedure is given in the last section of this chapter. For the present, CRAN comes to the rescue by providing a function, `boxTidwell`, contained in package `car`. Here is an example of this function's use.

First generate some data. For example, the code

```
x1<-runif(35,1,5);x2<-runif(35,0,1);x3<-runif(35,1,10)
y<-3-2*x1^2+sqrt(x2)-1/2*x3^(0.25)+rnorm(35,mean=0,sd=.1)
BoxTidData<-data.frame(y,x1,x2,x3); rm(y,x1,x2,x3)
```

uses the mean response function

$$E(y) = 3 - 2x_1^2 + \sqrt{x_2} - \frac{1}{2}\sqrt[4]{x_3}$$

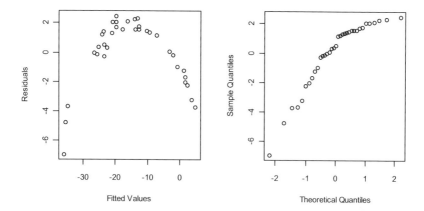

FIGURE 10.1: Plot of the residuals against fitted values and QQ normal probability plot of data that do not fit the traditional multiple regression model.

to produce data that most likely do not fit well to a model of the form

$$y_i = \beta_0 + \beta_1 x_{i1} + \beta_2 x_{i2} + \beta_3 x_{i3} + \varepsilon_i.$$

The data frame BoxTidData is stored in the file Data10x01.R.

To start, fit the data to an untransformed model and obtain some initial diagnostic plots[1] (see Figure 10.1) using

```
First.mod<-lm(y~x1+x2+x3,BoxTidData)
with(First.mod,plot(fitted.values,residuals,
        xlab="Fitted Values",ylab="Residuals"))
with(First.mod,qqnorm(residuals,main=NULL))
```

Figure 10.1 provides a quick initial graphical assessment and suggests that something is clearly very wrong with the proposed model.

To use the function boxTidwell, first load package car using the function library or require, and then execute

```
> with(BoxTidData,boxTidwell(y~x1+x2+x3))
        Score Statistic    p-value  MLE of lambda
x1         -44.074764   0.0000000      1.9952178
x2          -1.729491   0.0837213      0.2972401
x3           1.370600   0.1704998     -0.5597715
iterations = 5
```

---

[1] The functions win.graph and par were used to prepare a suitable graphics window.

The MLE of lambda column gives estimates of the powers for use in the transformations. The results are curious! One might expect the estimated powers to, at least, be of the same sign of the "true" powers — the power used for x3 to generate y was positive, the MLE of lambda computed for x3 is negative. It appears, also that in some cases (data dependent) the Box–Tidwell function runs into convergence problems, particularly if the mean square error, s^2, is large.

Putting these concerns aside and rounding the lambda estimates to one decimal place, choose $\hat{\lambda}_1 = 2$, $\hat{\lambda}_2 = 0.3$, and $\hat{\lambda}_3 = -0.6$, fit the data to the transformed model using

```
Second.mod<-lm(y~I(x1^2)+
    I(x2^(0.3))+I(x3^(-0.6)),BoxTidData)
```

and obtain diagnostic plots as was done for the first model (see Figure 10.2). Clearly good things have happened. Note, las=1 was used in both plots because the tick mark labels for the vertical axis appeared crowded.

There may be a question on the transformed model with regard to having x_1^2 in the model, but not x_1. If this is a concern, it is a simple matter to include the lower degree term, x_1, in the fitted model.

```
Third.mod<-lm(y~x1+I(x1^2)+
    I(x2^(0.3))+I(x3^(-0.6)),BoxTidData)
```

Diagnostic plots for Third.mod, such as in Figure 10.2, show little change and no adverse effects.

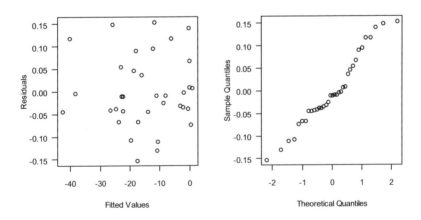

FIGURE 10.2: Diagnostic plots of residuals for transformed model with power estimates obtained using the Box–Tidwell procedure.

10.2.2 Transforming response and explanatory variables

The need for this scenario will typically arise if, initially, a normalizing or variance stabilizing transformation of the response variable has been performed. Data used in the following illustration were generated in a manner that produces right-skewed response variable data.

```
x1<-runif(35,0,1);x2<-runif(35,0,1)
y<-(2.3+1.25*x1^2-1.5*sqrt(x2)+rnorm(35,mean=0,sd=0.15))^2
XYTransData<-data.frame(y,x1,x2); rm(y,x1,x2)
```

The data frame `XYTransData` is contained in the file `Data10x02.R`. Quick diagnostics can be performed as before by first fitting the untransformed model

```
First.mod<-lm(y~x1+x2,XYTransData)
```

Figure 10.3 is produced using the earlier code. Again, something does not look right. Suppose it is known that the appropriate transformation for the response is the square-root transformation. Then, the fitted model

```
Second.mod<-lm(sqrt(y)~x1+x2,XYTransData)
```

shows some improvement as evidenced in Figure 10.4.

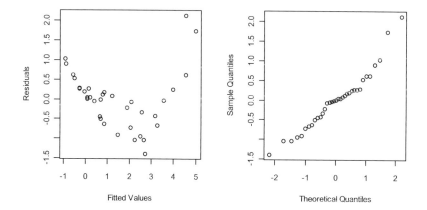

FIGURE 10.3: Diagnostic residual plots for the untransformed fitted model obtained from dataset `XYTransData`. Observe that the parameter `las=1` should probably have been used in these plots.

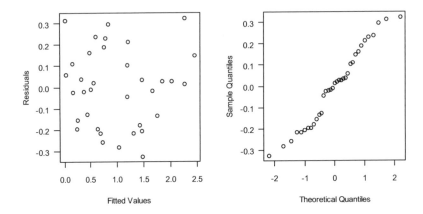

FIGURE 10.4: Diagnostic residual plots for the fitted model in which the observed responses were transformed using the square root transformation.

Further diagnostic plots shown in Figure 10.5 can be obtained using code of the form

```
attach(XYTransData)
with(Second.mod,plot(residuals~x1))
with(Second.mod,plot(residuals~x2))
plot(sqrt(y)~x1,ylab=expression(sqrt(y)))
plot(sqrt(y)~x2,ylab=expression(sqrt(y)))
detach(XYTransData)
```

The plots in Figure 10.4 do not provide any useful information with respect to further transformations. Moreover, plots of the residuals against the explanatory variables are of absolutely no help (see Figure 10.5). Similarly, while plots of the square root of the observed responses against the explanatory variables do suggest there might be a (very slight) need for transforming both explanatory variables, useful leads are not obvious. Now apply the Box–Tidwell procedure.

```
> with(XYTransData,boxTidwell(sqrt(y)~x1+x2))
        Score Statistic      p-value   MLE of lambda
x1            3.437036    0.0005881       2.0901015
x2            2.184379    0.0289344       0.7070977
iterations = 3
```

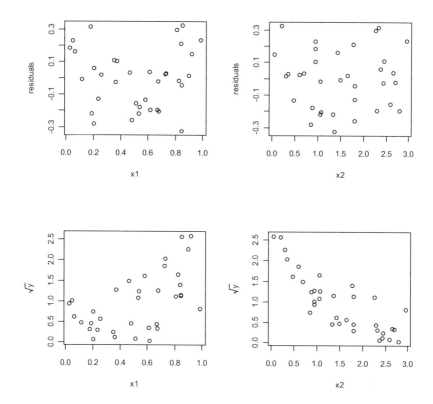

FIGURE 10.5: Diagnostic plots for the second model. While residual plots look acceptable, the plots of $\sqrt{y}$ against x_1 and x_2 *might* suggest a need for transformation of x_1 or x_2.

The estimated powers are $\hat{\lambda}_1 \approx 2.09$ and $\hat{\lambda}_2 \approx 0.71$. Since $\hat{\lambda}_1$ is close to 2 and $\hat{\lambda}_2$ is reasonably close to 0.5, it might be worthwhile fitting the data to the model[2]

$$\sqrt{y_i} = \beta_0 + \beta_1 x_{i1}^2 + \beta_2 \sqrt{x_{i2}} + \varepsilon_i.$$

The results show

```
> lm(sqrt(y)~I(x1^2)+sqrt(x2),XYTransData)$coefficients

(Intercept)    I(x1^2)    sqrt(x2)

  2.293927   1.082538   -1.452780
```

[2]Note that the question of whether to include x_1 in the model arises again. If x_1 is included in the proposed model, it will be found that it does not contribute significantly to the model. For this illustration x_1 is not included, right or wrong.

Observe that unlike the earlier example, here the Box–Tidwell procedure produced power estimates that are fairly close to the actual powers used to generate the data.

Before continuing, do not forget to detach package `car` and its accompanying packages.

10.3   Normalizing Transformations

All that applies to simple regression models applies here as well; see Section 7.3. For the current illustration, use the Box–Cox procedure to identify a power transformation for the dataset `XYTransData` from the previous example. To take advantage of the confidence interval feature produced by the function `boxcox`, Figure 10.6 was obtained using the code

```
library(MASS)
win.graph(width=6,height=3.5)
par(mfrow=c(1,2),ps=10,cex=.75)
boxcox(lm(y~x1+x2,XYTransData))
boxcox(lm(y~x1+x2,XYTransData),lambda=seq(.3,.55,.01))
par(mfrow=c(1,1)); detach(package:MASS)
```

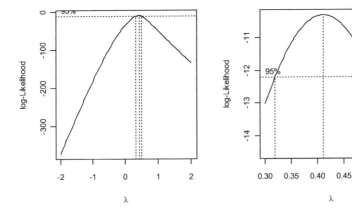

FIGURE 10.6: Plots obtained using function `boxcox` from package `MASS`. The right-hand plot is a zoomed-in version of the left.

This suggests using $\hat{\lambda} \approx 0.41$. However, since this value is closer to 0.5 than any other "nice" value and since 0.5 lies within the 95% confidence interval for λ, it would be reasonable to take a look at using the square root transformation. This would then lead back to the second phase of the previous example.

On a side note, it may be instructive to vary the value of sd (to larger and smaller values) in the data generating command used to get response values

```
y<-(2.3+1.25*x1^2-1.5*sqrt(x2)+rnorm(35,mean=0,sd=0.15))^2
```

and then repeat the Box–Tidwell *and* Box–Cox procedures to see how the estimated powers behave and vary.

10.4 Variance Stabilizing Transformations

For scenarios where variance stabilizing transformations work, the methods are as described in Section 7.4. For multiple linear regression models, however, there is a possiblity that alternative approaches will be needed. These methods are beyond the scope of this Companion; however, some pointers and places to look for help are given in Section 10.7 of this chapter.

10.5 Polynomial Regression

The goal is to estimate the true (but unknown) model with an mth degree polynomial model. The approach is as for simple regression, except that here products of two or more explanatory variables can show up.

Consider fitting the dataset BoxTidData to a second degree polynomial in x_1, x_2, and x_3.

```
Poly.mod<-lm(y~x1+x2+x3+I(x1*x2)+I(x1*x3)+
            I(x2*x3)+I(x1^2)+I(x2^2)+I(x3^2),BoxTidData)
```

Code for the residual plots in Figure 10.7 is similar to earlier code. The plots suggest reasonably well-behaved residuals. Out of curiosity try the step function out on Poly.mod.

```
> round(step(Poly.mod,trace=0)$coefficients,2)
(Intercept)    x2    x3  I(x2 * x3)  I(x1^2)  I(x2^2)  I(x3^2)
       2.63  1.85 -0.11      -0.04    -2.00    -0.70     0.01
```

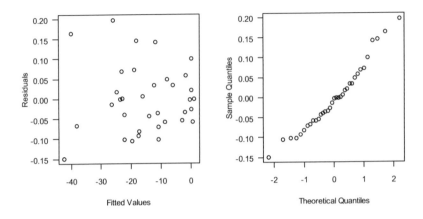

FIGURE 10.7: Diagnostic residual plots for the fitted second-degree poly-nomial model, `Poly.mod`, for dataset `BoxTidData`.

which suggests that an appropriate model for the data might be

$$y_i = \beta_0 + \beta_1 x_{i1} + \beta_2 x_{i2} + \beta_3 x_{i3} +$$
$$\beta_{23} x_{i2} x_{i3} + \beta_{11} x_{i1}^2 + \beta_{22} x_{i2}^2 + \beta_{33} x_{i3}^2 + \varepsilon_i.$$

Be sure to refer to the appropriate literature on interpreting and constructing polynomial regression models.

Now consider fitting the dataset `XYTransData` to a third-degree polynomial model.

```
Poly.mod2<-lm(sqrt(y)~x1+x2+I(x1^2)+I(x1*x2)+I(x2^2)
    +I(x1^3)+I(x1^2*x2)+I(x1*x2^2)+I(x2^3),XYTransData)
```

The diagnostic plots appear in Figure 10.8. Here

```
> round(step(Poly.mod2,trace=0)$coefficients,2)
(Intercept)    x2 I(x1^2) I(x2^2) I(x1^2 * x2) I(x1 * x2^2)
    2.36 -2.32    0.50    0.98         0.87        -0.27
I(x2^3)
  -0.15
```

suggests fitting the data `XYTransData` to a model of the form

$$\sqrt{y_i} = \beta_0 + \beta_1 x_{i1} + \beta_2 x_{i2} + \beta_{11} x_{i1}^2 + \beta_{22} x_{i2}^2 +$$
$$\beta_{112} x_{i1}^2 x_{i2} + \beta_{122} x_{i1} x_{i2}^2 + \beta_{222} x_{i2}^3 + \varepsilon_i.$$

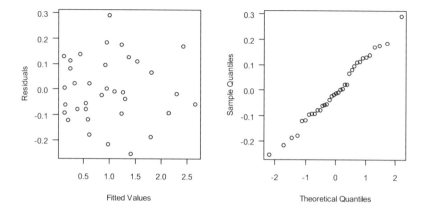

FIGURE 10.8: Diagnostic residual plots for the fitted third-degree polynomial model, `Poly.mod2`, for dataset `XYTransData`.

Comparing diagnostic plots for these two polynomial models against those of the models obtained using the Box–Tidwell procedure might provide some useful observations. However, comparing the models for purposes of selection is not straightforward since the two versions of models come under different classifications. Since nesting is absent, much of what was discussed in Chapter 9 does not apply. One option might be to use the PRESS statistic for each fitted model; however, be sure to refer to the literature to find out the accepted practice in such (non-nested) situations. There appear to be varying points of view.

10.6 Adding New Explanatory Variables

Although a little more is involved in multiple regression modeling, the ideas of including piecewise definitions and categorical variables as presented in Sections 7.6 and 7.7 still apply. Keep in mind that including an additional numerical explanatory variable in a simple regression model simply turns the model into a multiple regression model. Similarly, including an additional numerical explanatory variable in a multiple regression model simply increases the dimension of the domain space.

Thus, the possibilities include the following: Adding one or more "new" continuous explanatory variable to the proposed model; using subset or broken stick regression to partition the domain space; and partitioning the observed responses into categories by including categorical variables in the model. Since this process creates a natural nesting of models, the variable and model selection methods discussed in Sections 9.4 and 9.5 work well.

10.7   What if None of the Simple Remedies Help?

This can happen. If it does, there are a variety of alternative methods that may help address issues that cannot be satisfactorily resolved using the simple remedies discussed above, and in Chapter 7. Excluding issues with the normality assumption or the need for additional (continuous or categorical) explanatory variables, there are some extensions of the linear model as seen so far that include methods to address the types of issues mentioned below. The methods alluded to are beyond the scope of this Companion; however, details can typically be found in comprehensive texts. Specialized texts are available on specific topics, and some texts cover one or two such topics as supplements to fundamental ideas such as are addressed in this Companion. Further resources are listed in the bibliography of this Companion. As demonstrated in earlier portions of this Companion, if R functions are not available to perform a particular task, code can be written using resources from the literature.

10.7.1   Variance issues

Generalized least squares is a procedure that can be used when the error terms have unequal variances and are also correlated. The R function to construct a generalized least squares regression model is `gls` and is located in package `nlme` [55].

Weighted least squares is a special case of generalized least squares that can be used when the error terms have unequal variances and are not correlated. The function `gls` can be used to construct such models, too. Another approach is to introduce a vector of weights into the function `lm` by way of the argument `weights`. This instructs R to perform weighted least squares regression.

10.7.2   Outlier issues

When outliers pose (serious) problems, functions for *robust methods* such as `rlm`, which is contained in package `MASS`, `rq`, which is contained in package `quantreg` [44], and `lqs`, which is contained in package `MASS`, may help.

10.7.3   Multicollinearity issues

Principal components regression and *partial least squares regression* models can be constructed using the functions `pcr` and `plsr`, which are contained in package pls [70]. Additional functions for *principal components analysis* are available in package stats. The function lm.ridge, contained in package MASS, fits data to *ridge regression models*.

10.7.4 Autocorrelation issues

Autoregressive models can be fit to *time-series data* with the help of the functions ar and ar.ols, contained in package stats.

10.7.5 Discrete response matters

For *binary response* data, *logistic regression* can be performed using the function lrm in package Design [37]. This can also be performed using the versatile function glm in package stats. See also clogit in package survival. For other *discrete response* data, such as response with counts, *Poisson regression* can be performed using the function glm in package stats.

10.7.6 Distributional issues in general

These can usually be handled using the function glm in package stats. If not, there is a wealth of functions available for *nonparametric methods* in package stats and others; the help.search function with *suitable* keywords can be used to track down many of these, for example, help.search("kruskal") tracks down two items containing the keyword "kruskal." Remember, using several variations of keywords produces the broadest search results.

10.8 For the Curious: Box–Tidwell Revisited

The development of code to perform the Box–Tidwell procedure along the lines of the discussion in [53, pp. 192–193] is presented here. The dataset BoxTidData is used to provide a frame of reference with the earlier application of the function boxTidwell.

For the following discussion, let Y represent the response variable and X_j, $j = 1, 2, \ldots, p$, the explanatory variables. The goal is to obtain estimates for exponents $\lambda_1, \lambda_2, \ldots, \lambda_p$ for a transformed model having a mean response function of the form

$$E(y) = \beta_0 + \beta_1 w_1 + \cdots + \beta_p w_p,$$

where

$$w_j = \begin{cases} x_j^{\lambda_j} & \text{if } \lambda_j \neq 0, \\ \ln(x_j) & \text{if } \lambda_j = 0. \end{cases}$$

As shown in [53], if a Taylor series expansion about $(\lambda_1, \lambda_2, \ldots, \lambda_p)$ is applied to the proposed (transformed) mean response function and then only the linear part is considered, the result is

$$E(y) = \beta_0 + \beta_1 x_1 + \cdots + \beta_p x_p + \gamma_1 z_1 + \gamma_2 z_2 + \cdots + \gamma_p z_p, \tag{A}$$

where

$$\gamma_j = (\lambda_j - 1)\beta_j \quad \text{and} \quad z_j = x_j \ln(x_j) \quad \text{for} \quad j = 1, 2, \ldots, p. \tag{B}$$

The γ_j are regression parameters for the additional explanatory variables z_j. Note that λ_j is contained within the definition of γ_j.

To start things off, first fit the data to the model having the mean response function

$$E(y) = \beta_0 + \beta_1 x_1 + \cdots + \beta_p x_p,$$

and obtain the parameter estimates $\left(\hat{\beta}_1, \hat{\beta}_2, \cdots, \hat{\beta}_p\right)$.

Then fit the data to the model having the mean response function

$$E(y) = \beta_0^* + \beta_1^* x_1 + \cdots + \beta_p^* x_p + \gamma_1 z_1 + \gamma_2 z_2 + \cdots + \gamma_p z_p,$$

with z_j being as defined in (B), to obtain (only) the parameter estimates $\left(\hat{\gamma}_1, \hat{\gamma}_2, \cdots, \hat{\gamma}_p\right)$. The asterisks on the β_j^* are to differentiate these parameters from the previous model.

Next, using the relationship for γ_j in (B) and the parameter estimates $\left(\hat{\beta}_1, \hat{\beta}_2 \cdots, \hat{\beta}_p\right)$ from the previous model, compute

$$\hat{\lambda}_{1j} = \frac{\hat{\gamma}_j}{\hat{\beta}_j} + 1 \quad \text{for } j = 1, 2, \ldots, p.$$

Next, each x_j is replaced by

$$w_{1j} = \begin{cases} x_j^{\hat{\lambda}_{1j}} & \text{if } \hat{\lambda}_{1j} \neq 0, \\ \ln(x_j) & \text{if } \hat{\lambda}_{1j} = 0. \end{cases}$$

and the data are fitted to the transformed model with mean response function

$$E(y) = \beta_0 + \beta_1 w_{11} + \cdots + \beta_p w_{1p}. \tag{C}$$

Note, the iteration counter 1 in $\hat{\lambda}_{1j}$ and w_{1j} may or may not be needed in coding; it is included here for clarity.

The above steps are repeated if diagnostics procedures on the fitted model for (C) indicate a need, or if the magnitude of the difference between $(\hat{\lambda}_{11}, \hat{\lambda}_{12}, \ldots, \hat{\lambda}_{1p})$ and $(1, 1, \ldots, 1)$ is above a desired tolerance.

As an illustration, consider repeating the process a second time. Compute

$$z_j = w_{1j} \ln(w_{1j}) \quad \text{for} \quad j = 1, 2, \ldots, p,$$

then fit the data to the model with mean response function

$$E(y) = \beta_0^* + \beta_1^* w_{11} + \cdots + \beta_p^* w_{1p} + \gamma_1 z_1 + \gamma_2 z_2 + \cdots + \gamma_p z_p.$$

Then use the parameter estimates $\left(\hat{\beta}_1, \cdots, \hat{\beta}_p \right)$ from (C) and the estimates $(\hat{\gamma}_1, \hat{\gamma}_2, \cdots, \hat{\gamma}_p)$ from the above fitted model to compute the powers

$$\hat{\lambda}_{2j} = \frac{\hat{\gamma}_j}{\hat{\beta}_j} + 1 \quad \text{for } j = 1, 2, \ldots, p.$$

Some caution is appropriate here. The updated power estimates are obtained using the product $\hat{\lambda}_{1j} \hat{\lambda}_{2j}$, so from this second iteration, the updated transformation is given by

$$w_{2j} = \begin{cases} x_j^{\hat{\lambda}_{1j} \hat{\lambda}_{2j}} & \text{if } \hat{\lambda}_{1j} \hat{\lambda}_{2j} \neq 0, \\ \ln(x_j) & \text{if } \hat{\lambda}_{1j} \hat{\lambda}_{2j} = 0. \end{cases}$$

So, after k iterations of this process, the transformations would have the appearance

$$w_{2j} = \begin{cases} x_j^{\hat{\lambda}_{1j} \hat{\lambda}_{2j} \ldots \hat{\lambda}_{kj}} & \text{if } \hat{\lambda}_{1j} \hat{\lambda}_{2j} \ldots \hat{\lambda}_{kj} \neq 0, \\ \ln(x_j) & \text{if } \hat{\lambda}_{1j} \hat{\lambda}_{2j} \ldots \hat{\lambda}_{kj} = 0. \end{cases}$$

The question, then, is: When are the iterations stopped? In [53], it is suggested that the residual sum of squares be monitored. Another approach might be to measure the magnitude of the change in the lambda estimates.

Here is the sample code to perform the above steps on the dataset `BoxTidData`.

```
#Initialize the powers
10<-c(1,1,1)
#Define the transformation function w
w<-function(x,l){if (l!=0){return(x^l)
                 }else{return(log(x))}}
#Define the z function
z<-function(w){return(w*log(w))}
#Start initial step
mod0<-lm(y~I(w(x1,10[1]))+
```

```
        I(w(x2,10[2]))+I(w(x3,10[3])),BoxTidData)
#Extract parameter beta estimates
b<-mod0$coefficients[-1]
#Fit the "big" model
mod1<-lm(y~I(w(x1,10[1]))+I(w(x2,10[2]))+
        I(w(x3,10[3]))+I(z(w(x1,10[1])))+
          I(z(w(x2,10[2])))+I(z(w(x3,10[3]))),BoxTidData)
#Extract the gamma parameter estimates
g<-mod1$coefficients[5:7]
#Get updated lambdas
11<-10*(g/b+1)
#Fit the updated transformed model
mod3<-lm(y~I(w(x1,11[1]))+
        I(w(x2,11[2]))+I(w(x3,11[3])),BoxTidData)
#Compute the difference in lambdas, set loop counter
dif<-abs(sqrt(sum(10-11)^2)); k<-1
#If change in lambdas is above tolerance, begin loop
while ((k<=25)&(dif>.001)){
        #Get first stage model, betas, and lambdas
        mod0<-mod3; 10<-11; b<-mod0$coefficients[-1]
        #Fit the updated "big" model
        mod1<-lm(y~I(w(x1,10[1]))+I(w(x2,10[2]))+
                I(w(x3,10[3]))+I(z(w(x1,10[1])))+
                I(z(w(x2,10[2])))+I(z(w(x3,10[3]))),
                  BoxTidData)
        #Get new gammas
        g<-mod1$coefficients[5:7]
        #Update the lambdas
        11<-10*(g/b+1)
        #Fit the updated transformed model
        mod3<-lm(y~I(w(x1,11[1]))+I(w(x2,11[2]))+
                I(w(x3,11[3])),BoxTidData)
        #Compute difference in lambdas, set loop counter
        dif<-abs(sqrt(sum(10-11)^2)); k<-k+1}
```

```
#Print results
names(l1)<-c("Lambda 1","Lambda 2","Lambda 3")
l1;print(paste("Iterations = ",k-1,sep=""),quote=F)
```

The result is

```
  Lambda 1    Lambda 2     Lambda 3
 1.9952178   0.2972401   -0.5597715
[1] Iterations = 5
```

which matches up with the estimates obtained in Section 10.2.1 using the function boxTidwell.

Part III

Linear Models with Fixed-Effects Factors

Chapter 11

One-Factor Models

11.1 Introduction 251
11.2 Exploratory Data Analysis 253
11.3 Model Construction and Fit 254
11.4 Diagnostics 257
    11.4.1 The constant variance assumption 257
    11.4.2 The normality assumption 258
    11.4.3 The independence assumption 259
    11.4.4 The presence and influence of outliers 260
11.5 Pairwise Comparisons of Treatment Effects 261
    11.5.1 With a control treament 261
    11.5.2 Without a control treament 262
        11.5.2.1 Single pairwise t-test 262
        11.5.2.2 Simultaneous pairwise t-tests 263
        11.5.2.3 Tukey's HSD procedure 263
11.6 Testing General Contrasts 265
11.7 Alternative Variable Coding Schemes 267
    11.7.1 Treatment means model 267
    11.7.2 Treatment effects model 268
        11.7.2.1 Weighted mean condition 268
        11.7.2.2 Unweighted mean condition 270
    11.7.3 Diagnostics and pairwise comparisons 271
11.8 For the Curious 271
    11.8.1 Interval Estimates of Treatment Means 271
        11.8.1.1 t-intervals 272
        11.8.1.2 Scheffe F-intervals 272
    11.8.2 Tukey–Kramer pairwise tests 275
    11.8.3 Scheffe's pairwise F-tests 278
    11.8.4 Brown–Forsyth test 280
    11.8.5 Generating one-factor data to play with 281

11.1 Introduction

Consider a designed experiment in which observations for a response variable, Y, to a particular *factor* (a *categorical variable*) are grouped under p predetermined or fixed *factor levels* (or *treatments*). For the models under consideration in this chapter, the continuous response variable, Y, is viewed as being (potentially) dependent on the categorical explanatory variable defined by the factor. Let $j = 1, 2, \ldots, p$ denote treatment identifiers and let $i = 1, 2, \ldots, n_j$ identify the ith observed response within the jth treatment, where n_j represents the number of observations within the jth treatment;

allow for the possibility that the n_j differ across treatments. Denote the total number of observed responses by n and the ith observed response under the jth treatment by y_{ij}. Assuming random deviations exist in the observed responses, let ε_{ij} represent the random error term corresponding to the ith observed response within the jth treatment. There are two ways to look at things:

The *treatment* (or *cell*) *means model* has the appearance

$$y_{ij} = \mu_j + \varepsilon_{ij},$$

where the parameter μ_j represents the true *mean response under treatment j*.

Alternatively, let the parameter μ represent the unknown *overall mean response* and denote the unknown *mean treatment effect under treatment j* by the parameter τ_j. Using the decomposition $\mu_j = \mu + \tau_j$ in the treatment means model, the *treatment effects model* is given by

$$y_{ij} = \mu + \tau_j + \varepsilon_{ij}.$$

As for linear regression models, both models can be expressed in matrix form,

$$\mathbf{y} = \mathbf{X}\boldsymbol{\beta} + \boldsymbol{\varepsilon},$$

and it can be shown that the design matrix for the treatment means model has full column rank. The design matrix for the treatment effects model, however, does not have full column rank. A common remedy for this issue is to make use of an appropriate restriction on the effects parameters. The *unweighted mean condition* imposes the restriction $\sum_{j=1}^{p} \tau_j = 0$ and the *weighted mean condition* imposes the restriction $\sum_{j=1}^{p} w_j \tau_j = 0$, where the *weights*, w_j, satisfy $\sum_{j=1}^{p} w_j = 1$.

Alternatively, a coding scheme for the factor variable can be used that results in an equivalent, adjusted model for which the design matrix does have full column rank. This approach is the default scheme used by R.

Just as for linear regression models, there are analogous assumptions on the error terms. For all $j = 1, 2, \ldots, p$ and $i = 1, 2, \ldots, n_j$, it is assumed that the error terms are independent, and are identically and normally distributed with $\varepsilon_{ij} \sim N(0, \sigma^2)$.

If the categorical variable for the factor is denoted by Tr, data for such an experiment might have the appearance of the data shown in Table 11.1. The data are stored in the file `Data11x01.R` as a data frame named `OneWayData`. For future reference, when treatment sample (or *cell*) sizes are equal in magnitude, the experiment is referred to as having a *balanced design*; otherwise, the design is *unbalanced*. The R functions that perform the necessary computations address both designs (balanced or unbalanced) quite comfortably.

TABLE 11.1: Data for Chapter 11 Illustrations

Tr	Y												
1	2.2	3.1	1.5	3.1	2.8	2.6	2.1	2.1	2.5	3.4	2.5	2.6	1.8
2	2.8	3.0	2.6	3.8	2.6	3.1	3.1	3.1	3.0	2.9	3.4	3.3	3.0
3	2.4	1.9	1.6	1.9	3.0	2.0	3.5	1.9	2.3	3.4	2.5	2.9	3.0
4	3.1	3.6	1.6	3.7	3.5	1.8	3.3	4.1	3.3	3.7	4.1	3.9	
5	3.6	2.2	2.7	2.9	4.0								

Identifying treatment levels by numbers as in `OneWayData` can be convenient; however, it is important to make sure that the categorical variable `Tr` in `OneWayData` is defined as a factor. Do this using the `factor` function, for example, as follows:

```
OneWayData$Tr<-factor(OneWayData$Tr)
```

Treatment "values" can also be descriptive in form, such as names of rivers, etc. In such cases, while it is not necessary to use the `factor` function, there may be times when strange behavior in the computed results might suggest the function `factor` be used prior to performing computations.

11.2 Exploratory Data Analysis

Quick summaries of the data can be obtained using any one of the following commands:

```
summary(OneWayData)
by(OneWayData,OneWayData$Tr,summary)
with(OneWayData,tapply(y,Tr,summary))
```

The last two produce equivalent results and can be used to obtain a variety of details by simply replacing the function `summary` by functions like `mean`, `length`, and so on.

With respect to plots, there are times when two or more observed responses within a particular treatment might be equal or very close in value. In such situations, traditional plots might hide such cases — a single plotted point will show up for several cases. The `stripchart` function provides a means for addressing such cases. For example, Figure 11.1 was produced by executing the following code.[1]

---

[1] The functions `win.graph` and `par` were also used to prepare the graphics window.

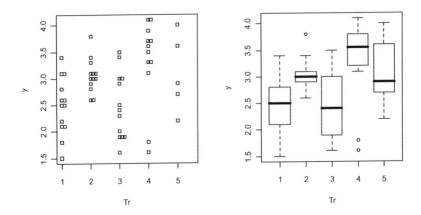

FIGURE 11.1: Stripchart with jittered points (left) and boxplots (right) of the observed responses in dataset `OneWayData`.

```
with(OneWayData,stripchart(y~Tr,
    method="stack",vertical=T,xlab="Tr",ylab="y"))
with(OneWayData,plot(y~Tr,ylab="y",xlab="Tr"))
```

Options for `method` in `stripchart` include `"stack"`, `"jitter"`, and `"overlay"`. The `stack` and `jitter` methods provide visual flags of observed response values that appear more than once for a treatment. Try `method="jitter"` and `method="overlay"` in `stripchart` for comparison.

Additional plotting functions that may find use include `hist` and `qqnorm`, which can be used to look at further distributional properties of the observed responses by treatments.

11.3   Model Construction and Fit

This first illustration looks at the treatment effects model using the default coding scheme in R. In this setting, the treatment effects model is reformulated as

$$y_{ij} = \begin{cases} \beta_0 + \varepsilon_{ij}, & \text{for } j = 1 \\ \beta_0 + \beta_j + \varepsilon_{ij}, & \text{for } j = 2, \ldots, p \end{cases}$$

where $i = 1, 2, \ldots, n_j$. It is important to remember that the original parameter interpretations do not apply to the parameters, β_j, in this setting. This approach, which essentially uses a model containing one less treatment parameter, is referred to as applying the *treatment coding* contrast in R (contr.treatment). If no changes are made to the *contrast* used to fit the model, R follows this approach by default and uses the first treatment as the *baseline treatment* (essentially sets the parameter corresponding to the first treatment equal to zero). It can be shown that the design matrix in the resulting adjusted model has full column rank.

More generally, the model fitting process begins with deciding upon a baseline treatment; this is more relevant to studies involving a control. If the treatments are arranged so that the first treatment is the desired baseline treatment, nothing needs to be done. However, if, for example, in a design with p treatments it is desired that the kth treatment be the baseline treatment, then the following sequence of commands are used:

> *contr* <-contr.treatment(p,base=k)
>
> *factor.list* <-C(*factor.list,contr*,how.many=p-1)

The function contr.treatment specifically addresses treatment coding, passing a matrix of contrasts into the object *contr*. This information is then passed into the factor list using the function C, which sets the contrasts' attribute for the factor. The contrasts' attribute then instructs R to use the information from *contr* to code the treatments appropriately, resulting in a model with a design matrix that has full column rank.

Finally, just to make sure that R uses the treatment coding scheme in constructing the model, particularly if the coding scheme needs to be reset, the desired scheme can be reset using the options function as follows:

> options(contrasts=c("contr.treatment","contr.poly"))

The function options can be used to set and examine a variety of global settings (or options). See the R documentation pages for C, contrasts, and options for details. These functions provide considerable power in performing quite a variety of tests through the function lm.

Going back to the dataset OneWayData, consider setting the contrast so that the last treatment is used as the baseline treatment. Then, under the treatment coding scheme, enter

> contr<-contr.treatment(5,base=5)
>
> OneWayData$Tr<-C(OneWayData$Tr,contr,how.many=4)
>
> options(contrasts=c("contr.treatment","contr.poly"))

Note that the options command is needed only if contrasts is not set on contr.treatment. This sets the stage for fitting the data to the model

$$y_{ij} = \begin{cases} \beta_0 + \varepsilon_{ij}, & \text{for } j = 5 \\ \beta_0 + \beta_j + \varepsilon_{ij}, & \text{for } j = 1,2,3,4. \end{cases}$$

Just as for linear regression models, the fitted model along with relevant summary statistics can be obtained using the function lm. For the present illustration, however, the following equivalent function is preferred:

```
tc.mod<-aov(y~Tr,OneWayData)
tc.sum<-summary.lm(tc.mod)
```

The function aov is referred to as an "lm wrapper" and is more suited for (purely) factorial studies.

The summary.lm function, when used on an aov object, produces details on specific parameter estimates in the lm format seen in earlier summaries for lm objects. This also produces familiar summaries and statistics needed for the diagnostics stage.

By comparing the models

$$y_{ij} = \mu + \tau_j + \varepsilon_{ij}$$

and

$$y_{ij} = \begin{cases} \beta_0 + \varepsilon_{ij}, & \text{for } j = p, \\ \beta_0 + \beta_j + \varepsilon_{ij}, & \text{for } j = 1, \ldots, p-1, \end{cases}$$

it can be shown that $\beta_0 = \mu + \tau_p$ and, for $k = 1, 2, \ldots, p-1$,

$$\beta_0 + \beta_k = \mu + \tau_k.$$

Combining these relationships with the (appropriate) treatment effects restriction results in the equivalent hypotheses

$$H_0 : \tau_1 = \tau_2 = \cdots = \tau_p = 0 \quad \Longleftrightarrow \quad H_0 : \beta_1 = \beta_2 = \cdots = \beta_{p-1} = 0.$$

Consequently, the model significance test seen in earlier models also assesses whether there is a significant difference in treatment effects, when applied to the treatment coded one-factor model. It is, however, convenient that the summary of an aov object produces the traditional ANOVA table.

```
> summary(tc.mod)
            Df  Sum Sq  Mean Sq  F value    Pr(>F)
Tr           4   6.796   1.6990   4.6131  0.002956 **
Residuals   51  18.783   0.3683
```

This provides the information needed to test the significance of treatment effects; that is,

$$H_0 : \tau_1 = \tau_2 = \cdots = \tau_5 = 0 \quad \text{vs.}$$
$$H_1 : \tau_j \neq 0 \text{ for at least one } j = 1, 2, \ldots, 5.$$

Information contained in the aov and summary.lm objects can be accessed as before to perform all diagnostic procedures appropriate for one-factor fixed-effects models.

11.4 Diagnostics

First extract the necessary statistics for use in the analyses. Since the explanatory variable is categorical, all that may be needed are the fitted values, residuals, studentized residuals, and possibly the studentized deleted residuals. The definitions and computational formulas for these remain as described in Section 8.4.

```
y.hat<-fitted.values(tc.mod)

e<-residuals(tc.mod)

r<-e/(tc.sum$sigma*sqrt(1-hatvalues(tc.mod)))

d<-rstudent(tc.mod)
```

A caution with respect to studentized residuals, which are denoted by r above, is appropriate here. If a particular treatment (or cell) has only one observed response, then the leverage corresponding to this case will be 1 and a value of Inf is assigned to the corresponding studentized residual; a value of NaN is assigned to the corresponding studentized deleted residual. In such cases, it is recommended that studentized and studentized deleted residuals not be used in any analyses.

Keep in mind that in any of the following plots, the identify function can be used to label suspicious or interesting points.

11.4.1 The constant variance assumption

Plots used to assess the validity of the constant variance assumption in linear regression also apply here. Since the explanatory variable is categorical, it is best to use jittered stripcharts. Figure 11.2 is the result of the following code:

```
plot(e~y.hat,xlab="Fitted values",ylab="Residuals")

stripchart(e~Tr,OneWayData,method="jitter",vertical=T,
                    ylab="Residuals",xlab="Treatments")
```

As with regression models, the plot of the residuals against the fitted values serves as a possible means of identifying transformations on the observed responses, if appropriate. Both plots in Figure 11.2 also serve as indicators of the possible presence of outliers.

While it would be a simple matter to alter the code for the earlier constructed function bf.Ftest, there is an added bonus for one-factor fixed-effects designs. The earlier mentioned built-in function hov, from the package HH, for the Brown–Forsyth test of the homogeneity of variances can be used:

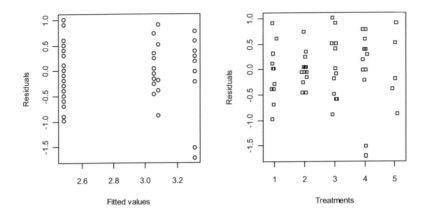

FIGURE 11.2: Diagnostic residual plots of the fitted model obtained from dataset `OneWayData`.

```
> hov(y~Tr,OneWayData)
        hov: Brown-Forsyth

data: y

F = 1.3025, df:Tr = 4, df:Residuals = 51, p-value = 0.2816

alternative hypothesis: variances are not identical
```

When done, be sure to remove HH and its companion packages in the same order as they were loaded by R.

11.4.2   The normality assumption

As with regression models, the QQ normal probability plot serves to assess whether the normality assumption on the error terms might be violated (see Figure 11.3). This plot also serves as an indicator of the possible presence of outliers.

```
qq<-qqnorm(r,main=NULL); with(qq,identify(x,y))
abline(a=0,b=1,lty=3)
```

Recall that if unstandardized residuals were to be used, the `qqline` should be used as the reference line.

Both the QQ normal probability correlation coefficient test and the Shapiro–Wilk test work here. For example, the function `qq.cortest` can be called up and used:

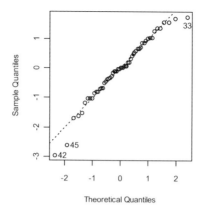

FIGURE 11.3: QQ normal probability plot of the studentized residuals with reference line $y = x$. The three extreme points were labeled using the `identify` function.

```
> qq.cortest(r,0.05)
      QQ Normal Probability Corr. Coeff. Test, alpha = 0.05
data: r, n = 56
RQ = 0.9857, RCrit = 0.979
alternative hypothesis: If RCrit > RQ, Normality assumption
is invalid
```

or

```
> shapiro.test(r)
      Shapiro-Wilk normality test
data: r
W = 0.9708, p-value = 0.1908
```

Both tests indicate that the assumption of normality need not be rejected.

11.4.3 The independence assumption

Earlier-mentioned methods used to look for a violation of the independence assumption apply, particularly if the data are obtained in a time-related sequence. These include any form of sequence plot that might aid in identifying auto- or spatial correlation. Time dependence and the absence of other important explanatory variables can also be assessed, *if* information on time and other variables in relation to the observed responses is available.

Scatter plots of the residuals against time and other variables can be obtained using the `plot` function as described earlier. The function `split` might also serve well in one-factor models when preparing plots that might help in identifying points in terms of new qualitative explanatory variables.

If applicable, the Durbin–Watson test can be conducted. Recall, this function is contained in package `lmtest` and the syntax is

```
dwtest(model)
```

where *model* is the name of the `aov` or `lm` object under study.

11.4.4   The presence and influence of outliers

With one-factor models, since the explanatory variable is qualitative, there is no need to assess issues relating to the explanatory variable. This leaves matters concerning the error terms, which reflect on the observed responses. For outlier analysis, any plot of the studentized deleted residuals (if available; see earlier caution) along with appropriate cutoff values serves the purpose. Consider looking at the outer 10% of the studentized deleted residuals for the current model.

Bonferroni cutoff values for the studentized deleted residuals are given by $\pm t(\alpha/(2m), n-p-1)$, where $m = 0.1n$ rounded up to the next integer. Figure 11.4, which contains horizontal lines representing Bonferroni cutoff lines, was obtained as follows. First, the basic plot is obtained using

```
plot(d~y.hat,ylim=c(-3.5,3.5),
    ylab=expression(hat(d)),xlab=expression(hat(y)))
```

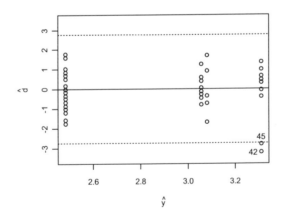

FIGURE 11.4: Plot of studentized deleted residuals against fitted values with Bonferroni cutoff values obtained for outer 10% ($m = 6$) cases.

Then, the cutoff lines are superimposed on the plot using

```
m<-ceiling(0.1*length(d)); df<-length(d)-5-1
a<-0.05/m; cv<-qt(a/2,df,lower.tail=F)
abline(h=c(-cv,0,cv),lty=c(3,1,3))
identify(y.hat,d)
```

Observe that two points are flagged as outliers.

The earlier-mentioned function `outlierTest` from package `car` can be used here. Also, some of the earlier-mentioned influence measures (using the function `influence.measures`) might be used as follows:

```
influence.measures(tc.mod)$is.inf[c(42,45),]
```

But it appears that influence analyses are typically not needed for (or performed on) one-factor fixed-effects models.

11.5 Pairwise Comparisons of Treatment Effects

Consider the options of having a control treatment and not having a control treatment.

11.5.1 With a control treament

For illustration purposes, suppose the last treatment is the control. Then, for $j = 1, 2, \ldots, p - 1$, it can be shown that

$$H_0 : \mu_j = \mu_p \quad \Longleftrightarrow \quad H_0 : \tau_j = \tau_p \quad \Longleftrightarrow \quad H_0 : \beta_j = 0.$$

These equivalences are nice, as they permit pairwise comparisons of treatment means and effects parameters using results on the parameters arising directly from the summary of the `aov` object with the treatment coding scheme. Analogous equivalences can be shown if the first treatment were to be assigned as the baseline treatment. For the illustration in question, the fifth treatment was taken as the control, and `tc.sum` contains

```
Coefficients:
```

	Estimate	Std. Error	t value	Pr(>|t|)	
(Intercept)	3.08000	0.27140	11.348	1.45e-15	***
Tr1	-0.59538	0.31936	-1.864	0.068	.
Tr2	-0.02615	0.31936	-0.082	0.935	
Tr3	-0.59538	0.31936	-1.864	0.068	.
Tr4	0.22833	0.32304	0.707	0.483	

Here, the line for `Tr1` represents $H_0 : \mu_1 = \mu_5$, the line for `Tr2` represents $H_0 : \mu_2 = \mu_5$, and so on.

11.5.2   Without a control treament

A variety of pairwise t-tests can be performed using the function `pairwise.t.test` from package `stats`, which has the general syntax

$$\texttt{pairwise.t.test}(\textit{Response},\textit{Factor},\texttt{p.adj="}\textit{choice}\texttt{")}$$

This function uses the original data to compare all pairs of parameters contained in the model, so pC_2 comparisons are performed.[2] If no adjustment is desired, use `p.adj="none"`. If an adjustment is desired for simultaneous comparisons, a variety of `p.adj` choices are available; here only the choice `"bonferroni"`, or `"bonf"` for short, is used. Since the variances are assumed equal, the default setting for pooled standard deviations is used.

For the computational details, suppose the jth and k^{th} treatment means (or treatment effects) parameters are to be compared. Then, since $\hat{\mu}_j - \hat{\mu}_k = \hat{\tau}_j - \hat{\tau}_k$, the test statistic for both the choices `"none"` and `"bonf"` is

$$t^* = \left| \hat{\mu}_j - \hat{\mu}_k \right| \bigg/ s \sqrt{\frac{1}{n_j} + \frac{1}{n_k}} \ .$$

If m denotes the number of simultaneous tests being performed, the p-value is computed using $P = m\,P(|t| > t^*)$ with $df = n - p$.

11.5.2.1   Single pairwise t-test

For one-at-a-time pairwise comparisons, use

```
> with(OneWayData,pairwise.t.test(y,Tr,p.adj="none"))

        Pairwise comparisons using t tests with pooled SD

data:  y and Tr
      1        2        3        4

2   0.0205    - - -

3   1.0000   0.0205     -            -

4   0.0014   0.2998   0.0014     -

5   0.0680   0.9351   0.0680   0.4829

P value adjustment method: none
```

The resulting matrix provides p-values for *least significant difference tests* (*LSD*) for all pairs.

---

[2] $pC_2 = p! / [2! \,(p-2)!]$, also referred to as the binomial coefficient, gives the number of ways two parameters can be chosen from p parameters. This is computed using `choose(p,2)`.

11.5.2.2 Simultaneous pairwise t-tests

Suppose m parameter pairs are to be compared simultaneously using the Bonferroni procedure. One approach is to simply use the output given previously by comparing the p-values in question against α/m. Alternatively, the function `pairwise.t.test` can be instructed to apply a Bonferroni adjustment to compare all possible pairs from the p treatments. If this is done, then while the test statistic remains as for the LSD test, the individual p-values in the function `pairwise.t.test` are computed using

$$P = \begin{cases} m\,P(|t| > t^*) & \text{if } m\,P(|t| > t^*) \le 1 \\ 1 & \text{if } m\,P(|t| > t^*) > 1 \end{cases}$$

where $df = n - p$. For simultaneous t-tests, the Bonferroni adjustment (or any other desired simultaneous test choice for `p.adj`) is passed into the function.

```
> with(OneWayData,pairwise.t.test(y,Tr,p.adj="bonf"))

        Pairwise comparisons using t tests with pooled SD

data: y and Tr

        1       2       3       4

2   0.205   -       -       -

3   1.000   0.205   -       -

4   0.014   1.000   0.014   -

5   0.680   1.000   0.680   1.000

P value adjustment method: bonferroni
```

Notice that the results here can be obtained by simply multiplying the LSD p-values by $5C_2 = 10$. If the resulting product is greater than or equal to one, a p-value of one is assigned.

11.5.2.3 Tukey's HSD procedure

When the number of pairwise comparisons is large, Bonferroni's procedure is less likely to reject null hypotheses that are false. Other procedures that are more suitable for such cases are available, one such procedure being *Tukey's HSD procedure*. This procedure uses the *studentized range distribution* to obtain a single critical value for the model in question regardless of the number of comparisons being performed. For a one-factor model, this procedure is executed using the general syntax

```
TukeyHSD(model,which,ordered=FALSE,conf.level=.95)
```

The arguments for this function are: *model*, the fitted linear model in question as an `aov` object (note that an `aov` object is required for `TukeyHSD`); and *which*, a character vector of the form `c("Tr1","Tr2",...)`. If this is not included, the default covers all pairwise comparisons. The argument setting `ordered=FALSE` is the default setting. By using `TRUE`, the factors are first arranged in ascending order according to cell sample means before calculating the differences; finally, the default confidence level is `conf.level=.95`.

For example:

```
> TukeyHSD(tc.mod,ordered=T)

  Tukey multiple comparisons of means

    95% family-wise confidence level

    factor levels have been ordered

Fit: aov(formula = y ~Tr, data = OneWayData)

$Tr
```

	diff	lwr	upr	p adj
1-3	4.440892e-16	-0.6731112	0.6731112	1.0000000
2-3	5.692308e-01	-0.1038804	1.2423420	0.1342564
5-3	5.953846e-01	-0.3076888	1.4984581	0.3493707
4-3	8.237179e-01	0.1367267	1.5107092	0.0113278
2-1	5.692308e-01	-0.1038804	1.2423420	0.1342564
5-1	5.953846e-01	-0.3076888	1.4984581	0.3493707
4-1	8.237179e-01	0.1367267	1.5107092	0.0113278
5-2	2.615385e-02	-0.8769196	0.9292273	0.9999894
4-2	2.544872e-01	-0.4325041	0.9414784	0.8319478
4-5	2.283333e-01	-0.6851326	1.1417992	0.9540102

A caution is appropriate here — be sure to look up how to interpret the results of Tukey's HSD to avoid contradictory results. It is also important to remember that the fitted model used *must include the intercept term*, that is, `"-1"` *must not* be placed in the model formula for `aov`; see Section 11.7.1.

Another point to keep in mind is that while Tukey's HSD procedure applies only to balanced designs, it is stated in the R documentation for `TukeyHSD` that the function incorporates an adjustment for unequal cell sizes that produces acceptable results for mildly unbalanced designs.

Sample code for the *Tukey–Kramer procedure*, which works on balanced and unbalanced designs, is demonstrated in the last section of this chapter.

11.6 Testing General Contrasts

Pairwise tests are simple cases of what are referred to as *contrasts*,[3] the null hypotheses of which have the general form

$$H_0 : \sum_{j=1}^{p} c_j \mu_j = 0 \quad \text{or} \quad H_0 : \sum_{j=1}^{p} c_j \tau_j = 0,$$

where $\sum_{j=1}^{p} c_j = 0$. It can be observed that performing a general contrast is somewhat analogous to a partial F-test. Suppose

$$H_0 : \mathbf{C}\boldsymbol{\beta} = \mathbf{0} \quad \text{vs.} \quad H_1 : \mathbf{C}\boldsymbol{\beta} \neq \mathbf{0}$$

is to be tested, where

$$\mathbf{C} = \begin{bmatrix} c_{11} & c_{12} & \cdots & c_{1p} \\ c_{21} & c_{22} & \cdots & c_{2p} \\ \vdots & \vdots & \ddots & \vdots \\ c_{m1} & c_{m2} & \cdots & c_{mp} \end{bmatrix}$$

with $\sum_{j=1}^{p} c_{ij} = 0$ for each $i = 1, 2, \ldots, m$. The matrix $\mathbf{C}$ describes the m contrasts to be tested simultaneously. Suppose $\mathbf{C}$ has $q \leq m$ linearly independent rows. Then the test statistic associated with *this* general contrast is

$$F_{\mathbf{C}}^* = \frac{\left(\mathbf{C}\hat{\boldsymbol{\beta}}\right)' \left[\mathbf{C}\left(\mathbf{X}'\mathbf{X}\right)^{-1}\mathbf{C}'\right]^{-1}\left(\mathbf{C}\hat{\boldsymbol{\beta}}\right)}{q\,s^2},$$

and the p-value is computed using $P = P(F > F_{\mathbf{C}}^*)$ with $df_N = q$ and $df_D = n - p$. Package `multcomp` [41] contains functions that perform the needed computations for testing such contrasts. Here is a general description of the process.

A *general linear hypothesis test* object is created, the basic function call of which is

```
ghlt(model,linfct=mcp(factor=contr))
```

where *model* is the aov (or lm) object of interest, *factor* is the name of the factor variable, and *contr* is the desired contrast or set of contrasts to be tested. Results of the test can then be obtained using the ever useful `summary` function. Here is a simple example.

---

[3] Be aware, this does not necessarily mean the same thing as it does with reference to model fitting functions in R.

For the dataset `OneWayData`, consider testing the hypotheses

$$H_0 : \mu_1 - \tfrac{1}{4}\mu_2 - \tfrac{1}{4}\mu_3 - \tfrac{1}{4}\mu_4 - \tfrac{1}{4}\mu_5 = 0 \qquad \text{vs.}$$
$$H_1 : \mu_1 - \tfrac{1}{4}\mu_2 - \tfrac{1}{4}\mu_3 - \tfrac{1}{4}\mu_4 - \tfrac{1}{4}\mu_5 \neq 0$$

First create the contrast matrix,

```
contr<-rbind("m1-.25m2-.25m3-.25m4-.25m5"=
                    c(1,-.25,-.25,-.25,-.25))
```

then create the general linear hypothesis test object and obtain its summary.

```
library(multcomp)
glht.object<-glht(tc.mod,linfct=mcp(Tr=contr))
summary(glht.object,Ftest())
```

The argument `Ftest()` in the summary function instructs R to provide summary results of the global (joint) F-test for the (full) contrast. The output shows

```
            General Linear Hypotheses

   Multiple Comparisons of Means: User-defined Contrasts

   Linear Hypotheses:
                                           Estimate
   m1-.25m2-.25m3-.25m4-.25m5 == 0    -0.4971
   Global Test:
            F DF1 DF2  Pr(>F)
   1 6.436   1   51 0.01429
```

If the argument `Ftest()` is left out of the `summary` command, a set of simultaneous t-tests (called Scheffe's simultaneous t-tests by some) for each row of the contrast matrix is performed. For the current example, the output contains test results for the single contrast defined. It will also be observed that the square of the t-value in the simultaneous test output equals the F-value for the global F-test above; this occurs only for single contrast general linear hypothesis tests. If the contrast matrix had more than one row, the same number of simultaneous t-tests would be performed, one for each contrast.

The above process does not provide a direct means of performing Scheffe's simultaneous F-tests. The simultaneous t-tests are actually equivalent to a sequence of partial F-tests performed on the tweaked model (resulting from the application of the contrast matrix).

When done, do not forget to detach package `multcomp` and its accompanying packages in the same order as they were loaded.

11.7 Alternative Variable Coding Schemes

Three alternatives to the treatment coding scheme are shown here. The discussion begins with testing the significance of treatment effects for each of the three alternative schemes. This is then followed by a brief discussion of diagnostics and pairwise comparisons, the methods of which turn out to be identical to the default setting.

11.7.1 Treatment means model

The simplest alternative is to use the *treatment* (or *cell*) *means model*, $y_{ij} = \mu_j + \varepsilon_{ij}$, where $j = 1, 2, \ldots, p$ and $i = 1, 2, \ldots n_j$ and the treatment sample sizes, n_j, need not be equal. The structural setup for this model is such that the design matrix does not have a first column containing 1's and has full column rank; so, there is no need to adjust the model. Computationally, the process is analogous to linear regression without the intercept term. The fitting command for this model is as before, with one small tweak:

```
tm.mod<-aov(y~Tr-1,OneWayData)

tm.sum<-summary.lm(tm.mod)
```

The "-1" in the formula argument instructs R *not to include the constant term in the model.* In this case, the coefficients in tm.mod are the estimated treatment means, $\hat{\mu}_1$, $\hat{\mu}_2$, $\hat{\mu}_3$, $\hat{\mu}_4$, and $\hat{\mu}_5$. The temptation might then be to obtain

```
> summary(tm.mod)
             Df Sum Sq  Mean Sq F value    Pr(>F)
Tr            5 460.52   92.103  250.08  < 2.2e-16
Residuals    51  18.78    0.368
```

However, be aware that the model significance test *has not been performed*, see Section 11.3. In this situation, the function aov tests the hypotheses

$H_0 : \mu_1 = \mu_2 = \cdots = \mu_p = 0$ vs.

$H_1 : \mu_j \neq 0$ for at least j.

where the null hypothesis corresponds to the reduced model $y_{ij} = \varepsilon_{ij}$. To test the significance of treatment effects (or, equivalently, the difference in treatment means), follow the partial F-test approach. First construct the reduced model $y_{ij} = \mu + \varepsilon_{ij}$, then compare it with the full model $y_{ij} = \mu_j + \varepsilon_{ij}$ using the function anova:

```
> red.model<-aov(y~1,OneWayData)
> anova(red.model,tm.mod)
Analysis of Variance Table
Model 1: y ~1
Model 2: y ~Tr - 1
  Res.Df     RSS  Df  Sum of Sq       F    Pr(>F)
1     55  25.579
2     51  18.783   4      6.796  4.6131  0.002956
```

These results match earlier results obtained using the treatment coding scheme.

11.7.2 Treatment effects model

Recall the *treatment effects model*, $y_{ij} = \mu + \tau_j + \varepsilon_{ij}$, where $\mu_j = \mu + \tau_j$. The design matrix for this model does not have full column rank and, therefore, the model needs adjusting to an equivalent full rank model. There are a couple of options.

11.7.2.1 Weighted mean condition

If a *weighted mean*, $\mu = \sum_{j=1}^{p} w_j \mu_j$, is assumed, where the *weights* w_j satisfy $\sum_{j=1}^{p} w_j = 1$, then the relation $\mu_j = \mu + \tau_j$ gives rise to a *weighted sum contrast*, $\sum_{j=1}^{p} w_j \tau_j = 0$. One way to assign weights is to use treatment sample sizes, in which case $w_j = n_j/n$. This provides a method for obtaining an equivalent full rank model.

For $j = 1, 2, \ldots, p$, and each $i = 1, 2, \ldots, n_j$, let x_{ij} be the ij^{th} entry in the design matrix of the adjusted model. It can be shown that the above condition results in

$$x_{ij} = \begin{cases} 1 & \text{for } y_{ij}, \\ 0 & \text{for } y_{ik} \text{ where } k \neq j \text{ and } k \neq p \\ -a_j & \text{for } y_{ip}. \end{cases}$$

where $a_j = n_j/n_p$ for $j = 1, 2, \ldots, p-1$, and the design matrix for this adjusted model has full column rank. Furthermore, it can be shown that the "new" (adjusted) design matrix for the model can be obtained by left-multiplying the "original" (unadjusted) design matrix by

$$\begin{bmatrix} 1 & 0 & 0 & \cdots & 0 \\ 0 & 1 & 0 & \cdots & 0 \\ 0 & 0 & 1 & \cdots & 0 \\ \vdots & \vdots & \vdots & \ddots & \vdots \\ 0 & 0 & 0 & \cdots & 1 \\ 0 & -a_1 & -a_2 & \cdots & -a_{p-1} \end{bmatrix}$$

The desired "contrast," call this **K**, for use by R, is then obtained by removing the first row and column from this matrix; thus,

$$\mathbf{K} = \begin{bmatrix} 1 & 0 & \cdots & 0 \\ 0 & 1 & \cdots & 0 \\ \vdots & \vdots & \ddots & \vdots \\ 0 & 0 & \cdots & 1 \\ -a_1 & -a_2 & \cdots & -a_{p-1} \end{bmatrix}$$

Notice that the number of rows in **K** equals the number of factor levels in the unadjusted model, and the number of columns equals the number of (nonredundant) factor levels in the adjusted model.

Setting user defined contrasts in R may be accomplished as follows. First create the matrix **K**:

```
a<-numeric(4)
attach(OneWayData)
for (i in 1:4) {a[i]<-length(Tr[Tr==i])/length(Tr[Tr==5])}
detach(OneWayData)
K<-matrix(c(1,0,0,0,-a[1],
            0,1,0,0,-a[2],
            0,0,1,0,-a[3],
            0,0,0,1,-a[4]),nrow=5,ncol=4)
```

Then use the function C to set the contrast

```
OneWayData$Tr<-C(OneWayData$Tr,K,how.many=4)
```

The argument `how.many` is to indicate how many nonredundant levels remain. This is the number of effects parameters minus the number of redundent effects parameters, which is 1 in the one-factor case. Once this is done, the treatment effects model is fitted using the function aov.

```
wm.mod<-aov(y~Tr,OneWayData)
wm.sum<-summary.lm(wm.mod)
```

In this case, the `coefficients` in `wm.mod` are the parameter estimates $\hat{\mu}$, $\hat{\tau}_1$, $\hat{\tau}_2$, $\hat{\tau}_3$, and $\hat{\tau}_4$. The remaining parameter estimate, $\hat{\tau}_5$, can then be computed using $\hat{\tau}_5 = -\sum_{j=1}^{4} a_j \hat{\tau}_j$, where $a_j = n_j/n_5$. The residuals, fitted values and other model statistics can be accessed from `wm.mod` and `wm.sum`. For example, to test the hypotheses

$$H_0 : \tau_1 = \tau_2 = \cdots = \tau_5 = 0 \quad \text{vs.}$$
$$H_1 : \tau_j \neq 0 \text{ for at least one } j = 1, 2, \ldots, 5,$$

simply enter

```
summary(wm.mod)
```

to get results that are identical to the significance test for the earlier two cases; there is no need to compare a reduced and a full model here.

Caution: Be sure to reset the contrasts appropriately if a different set of data is to be fitted to a model. A simple way to do this is to restart R.

11.7.2.2   Unweighted mean condition

If the data were to be balanced, the weighted mean case reduces to the built-in *sum contrast* in R. An *unweighted mean* assumption, $\mu = \sum_{j=1}^{p} \mu_j / p$, gives rise to the (*unweighted*) *sum contrast* $\sum_{j=1}^{p} \tau_j = 0$ for balanced designs and, more generally, $\sum_{j=1}^{p} \beta_j = 0$ for a model of the form

$$y_{ij} = \beta_0 + \beta_j + \varepsilon_{ij}.$$

Note that in the case of balanced data, everything about the weighted sum model and the sum contrast model is identical. In the case of unbalanced data, parameter interpretations and estimates may differ. As with the weighted mean case, it can be shown that the desired contrast, **K**, for use by R is

$$\mathbf{K} = \begin{bmatrix} 1 & 0 & \cdots & 0 \\ 0 & 1 & \cdots & 0 \\ \vdots & \vdots & \ddots & \vdots \\ 0 & 0 & \cdots & 1 \\ -1 & -1 & \cdots & -1 \end{bmatrix}$$

Once again, the number of rows in **K** equals the number of factor levels in the unadjusted model, and the number of columns equals the number of (nonredundant) factor levels in the adjusted model. Code equivalent to that presented in the previous example can be used to prepare the matrix **K**; however, it is simpler to use a built-in feature of R:

```
K<-contr.sum(5)
OneWayData$Tr<-C(OneWayData$Tr,K,how.many=4)
sc.mod<-aov(y~Tr,OneWayData)
sc.sum<-summary.lm(sc.mod)
```

sets the *sum contrasts* and obtains the relevant model and model summary objects. The number 5 in `contr.sum` indicates the number of treatments in the unadjusted model. The computed parameter estimates in `sc.mod` differ from those obtained using the weighted sum contrast; however,

```
summary(sc.mod)
```

produces results that are identical to the significance tests for all previous settings.

11.7.3 Diagnostics and pairwise comparisons

The fitted values, residual standard error, and residuals for all the alternatives discussed are identical (allowing for roundoff differences) to those of the treatment coded model. These can be obtained as before using commands of the form

```
y.hat<-fitted.values(model)
e<-residuals(model)
r<-e/(model$sigma*(1-hatvalues(model)))
d<-rstudent(model)
```

Consequently, the diagnostics for the four models are identical.

With respect to pairwise comparisons of means or effects parameters, the R function `pairwise.t.test` uses the original data to perform the tests and, consequently, tests treatment means (and effects) directly. Also, `TukeyHSD` produces identical results. So, for pairwise t-tests and Tukey's HSD, use the commands

```
with(data,pairwise.t.test(y,Tr,p.adj="none"))
with(data,pairwise.t.test(y,Tr,p.adj="bonf"))
TukeyHSD(model,ordered=T)
```

Keep in mind that `TukeyHSD` does not like the treatment means model; however, it does produce identical results for the other three models.

In performing general contrasts, the function `glht` appears to produce identical results for all four models.

11.8 For the Curious

This section looks at obtaining interval estimates of treatment means, a couple of programming exercises in performing simultaneous pairwise comparisons using the Tukey–Kramer and Sheffe proceedures. Also, the earlier constructed Brown–Forsyth test function is revisited, and a data generating example is given.

11.8.1 Interval Estimates of Treatment Means

Two types are considered: t-intervals, which cover one-at-a-time estimates, and Bonferroni simultaneous intervals. Next, Scheffe intervals might be preferred if there is a large number of treatment means to be estimated.

11.8.1.1   t-intervals

The formulas to obtain t-intervals for treatment means are essentially the same as those for regression models; parameter and estimate labels look different but all else is equivalent,

$$\hat{\mu}_j - t\left(\alpha/\left(2m\right), n - p\right) s_{\hat{\mu}_j} < \mu_j < \hat{\mu}_j + t\left(\alpha/\left(2m\right), n - p\right) s_{\hat{\mu}_j}.$$

Consequently, if the treatment means model `tm.mod` is used, the function `confint` used earlier works here, too. If $m = 1$, the one-at-a-time intervals are computed and if $m > 1$, then m simultaneous intervals are computed. For example,

```
> confint(tm.mod)
         2.5 %      97.5 %
Tr1   2.146704   2.822527
Tr2   2.715935   3.391758
Tr3   2.146704   2.822527
Tr4   2.956624   3.660043
Tr5   2.535134   3.624866
```

gives the 95% one-at-a-time intervals for μ_1, μ_2, μ_3, μ_4, and μ_5. On the other hand,

```
> confint(tm.mod,level=1-.05/5)
         0.5 %      99.5 %
Tr1   2.034244   2.934986
Tr2   2.603475   3.504217
Tr3   2.034244   2.934986
Tr4   2.839572   3.777094
Tr5   2.353799   3.806201
```

produces the 95% simultaneous (Bonferroni) intervals for μ_1, μ_2, μ_3, μ_4, and μ_5.

11.8.1.2   Scheffe F-intervals

Simultaneous F-intervals for all the parameters using Scheffe's procedure can be obtained if there is concern about the number of parameters being estimated simultaneously. Consider using the treatment coded model and suppose the last parameter is treated as the redundant parameter. Then the adjusted model has the appearance

$$y_{ij} = \beta_0 + \beta_j + \varepsilon_{ij},$$

with $\beta_p = 0$, where $j = 1, 2, \ldots, p$ and $i = 1, 2, \ldots, n_j$. Comparing with the treatment means model,

$$y_{ij} = \mu_j + \varepsilon_{ij}, \quad j = 1, 2, \ldots, p,$$

it can be shown that

$$\hat{\mu}_p = \hat{\beta}_0 \quad \text{and} \quad \hat{\mu}_j = \hat{\beta}_0 + \hat{\beta}_j \quad \text{for } j = 1, 2, \ldots, p - 1$$

To obtain interval estimates for treatment means parameters, it is necessary to obtain the estimated means as well as

$$
\begin{aligned}
s_{\hat{\mu}_p} &= s_{\hat{\beta}_0} \quad \text{and} \\
s_{\hat{\mu}_j} &= s_{\hat{\beta}_0 + \hat{\beta}_j} \\
&= \sqrt{\operatorname{var}(\hat{\beta}_0 + \hat{\beta}_j)} \\
&= \sqrt{\operatorname{var}(\hat{\beta}_0) + 2 \ \operatorname{cov}(\hat{\beta}_0, \hat{\beta}_j) + \operatorname{var}(\hat{\beta}_j)}.
\end{aligned}
$$

Estimates for the variance and covariance terms are contained in $s^2 (\mathbf{X'X})^{-1}$. These values are then placed in the formula

$$\hat{\mu}_j - \sqrt{p\,F\,(\alpha, p, n - p)}\, s_{\hat{\mu}_j} < \mu_j < \hat{\mu}_j + \sqrt{p\,F\,(\alpha, p, n - p)}\, s_{\hat{\mu}_j},$$

with $df_N = p$ and $df_D = n - p$. The object tc.sum contains s as sigma and $(\mathbf{X'X})^{-1}$ as cov.unscaled; df_N and df_D are contained in df.

Consider obtaining 95% Scheffe intervals for OneWayData when the default treatment coded model, tc.mod, is in use. Note: remove all objects in the workspace except OneWayData, tc.mod and tc.sum. Keeping in mind that all relevant statistics are contained in tc.sum, the following code[4] performs the above calculations.

```
#Attach tc.sum for convenience
attach(tc.sum)
#Initialize storage locations
mus<-numeric(5);S.ers<-numeric(5)
lwr<-numeric(5); upr<-numeric(5)
#List names of entries
labs<-c("mu1","mu2","mu3","mu4","mu5")
#Get degrees of freedom for F-statistic
df1<-df[1];df2<-df[2]
```

---

[4]Problems may arise when the object tc.sum is attached. Try executing the code before and after the recommended "cleanup." See what happens.

```
#Calculate the "critical value"
cv<-sqrt(df1*qf(.05,df1,df2,lower.tail=F))
#Use for-loop to get statistics for first four parameters
for (i in 1:4) {
    #Get estimated treatment means
    mus[i]<-coefficients[1,1]+coefficients[i+1,1]
    #Get corresponding standard errors, close for-loop
    S.ers[i]<-sigma*sqrt(cov.unscaled[1,1]+
        cov.unscaled[i+1,i+1]+2*cov.unscaled[1,i+1])}
#Get fifth sample mean and corresponding standard error
mus[5]<-coefficients[1,1]
S.ers[5]<-sigma*sqrt(cov.unscaled[1,1])
detach(tc.sum)
#Start loop to get interval estimates
for (j in 1:5){
    lwr[j]<-round(mus[j]-cv*S.ers[j],4)
    upr[j]<-round(mus[j]+cv*S.ers[j],4)}
#Store results
F.Intervals<-data.frame("Est."=mus,"S.er"=S.ers,
                "Lower"=lwr,"Upper"=upr,row.names=labs)
```

The contents of F.Intervals are

	Est.	S.er	Lower	Upper
mu1	2.484615	0.1683176	1.9020	3.0673
mu2	3.053846	0.1683176	2.4712	3.6365
mu3	2.484615	0.1683176	1.9020	3.0673
mu4	3.308333	0.1751905	2.7019	3.9148
mu5	3.080000	0.2714039	2.1405	4.0195

This code could very easily be turned into a function; however, an effort would have to be made to make it "foolproof" and usable on any model. Once again, remove all objects except OneWayData, tc.mod, and tc.sum.

11.8.2 Tukey–Kramer pairwise tests

For the sheer joy of the exercise, a bare-bones program to perform Tukey–Kramer pairwise comparisons can be written. The nuts and bolts of this procedure can be found in the literature; for example, see [45, pp. 746–752].

Two new functions are needed to compute probablities and quantiles for the *studentized range distribution*. Default arguments that need not be changed have not been included.

```
ptukey(q.Stat, p, n-p, lower.tail = FALSE)

qtukey(alpha, p, n-p, lower.tail = FALSE)
```

Steps to perform the various computations for the Tukey–Kramer procedure may be summarized as follows:

1. Obtain and arrange the sample means in ascending order to get $\hat{\mu}_{(1)}$, $\hat{\mu}_{(2)}$, $\hat{\mu}_{(3)}$, ..., $\hat{\mu}_{(p)}$. The corresponding sample sizes are denoted by $n_{(1)}, n_{(2)}, \ldots, n_{(p)}$. The parenthesized subscripts are used to indicate a reordered list.

2. Let s denote the residual standard error and, for $k = p, (p-1), \ldots, 2$, conduct the first test sequence using the differences $\mu_{(k)} - \mu_{(1)}$ on the full set $\hat{\mu}_{(1)}, \hat{\mu}_{(2)}, \hat{\mu}_{(3)}, \ldots, \hat{\mu}_{(p)}$ as follows:

   (a) Compute upper and lower bounds for confidence intervals using

   $$\hat{\mu}_{(k)} - \hat{\mu}_{(1)} \pm q(\alpha, p, n-p) \, s \sqrt{\frac{1}{2}\left(\frac{1}{n_{(k)}} + \frac{1}{n_{(1)}}\right)}.$$

   (b) Compute the test statistics (or q-statistics),

   $$q^* = \frac{\hat{\mu}_{(k)} - \hat{\mu}_{(1)}}{s \sqrt{\frac{1}{2}\left(\frac{1}{n_{(k)}} + \frac{1}{n_{(1)}}\right)}}.$$

   By arranging the parameter estimates in ascending order, the test statistics are guaranteed to be positive.

   (c) p-values for each test statistic, q^*, are computed as for a right-tailed test using $P(q > q^*)$.

3. Subsequent test sequences are performed by dropping the lowest term and repeating the above process. For example, for $k = p, (p-1), \ldots, 3$, the second test sequence would be on the differences $\mu_{(k)} - \mu_{(2)}$ and would involve $\hat{\mu}_{(2)}, \hat{\mu}_{(3)}, \ldots, \hat{\mu}_{(p)}$. Formulas for the interval estimates and q-statistics would be analogous to those described in Step 2.

4. The process is repeated until the last comparison $\mu_{(p)} - \mu_{(p-1)}$.

The following code produces the details laid out previously. The sample means are obtained from the raw data and the residual standard error can be obtained from the `summary.lm` object of any one of the models discussed previously, `tc.sum` is used here.

```
#Get sample means and sizes
mus<-with(OneWayData,tapply(y,Tr,mean))
ns<-with(OneWayData,tapply(y,Tr,length))
#Get ordering, number of samples, and entry names
pos<-order(mus); p<-length(mus)
labs<-c("mu1","mu2","mu3","mu4","mu5")
#Store this stuff and clean house a bit
stuff<-data.frame(cbind(pos,mus,ns),row.names=labs)
rm(mus,pos,ns)
#Initialize comparison counter, get number of comparisons
k<-1; m<-choose(5,2)
#Get standard error and degrees of freedom
s<-tc.sum$sigma; dfs<-tc.sum$df[2]
#Get "q-critical value" for interval estimates, use alpha = 0.05
q<-qtukey(.05,p,dfs,lower.tail=F)/sqrt(2)
#Initialize the various storage spaces
pair<-c(rep(" ",m)); diff<-numeric(m);
lwr<-numeric(m); upr<-numeric(m)
q.stat<-numeric(m); p.value<-numeric(m)
#Attach stuff for convenience
attach(stuff)
#Begin outer (ascending) for-loop
for (i in 1:4){
    #Begin inner(descending) for-loop
    for (j in seq(5,i+1,-1)){
        #Get names of pairs being compared
        pair[k]<-paste(row.names(stuff[pos==j,]),
            "-",row.names(stuff[pos==i,]))
        #Compute difference between sample means
        diff[k]<-round(mus[pos==j]-mus[pos==i],4)
        #Compute lower bound for interval
```

```
    lwr[k]<-round(diff[k]-
        q*s*sqrt(1/ns[pos==j]+1/ns[pos==i]),4)
    #Compute upper bound for interval
    upr[k]<-round(diff[k]+
        q*s*sqrt(1/ns[pos==j]+1/ns[pos==i]),4)
    #Get q-statistic
    q.stat[k]<-round(diff[k]/
        (s*sqrt((1/ns[pos==j]+1/ns[pos==i])/2)),4)
    #Get p-value
    p.value[k]<-round(ptukey(q.stat[k],
        p,dfs,lower.tail=F),4)
    #Update comparison counter and close both loops
    k<-k+1}}
detach(stuff)
#Store details
TK.Comps<-data.frame("Pairs"=pair,"Diff"=diff,"Lower"=lwr,
    "Upper"=upr,"q-Stat"=q.stat,"p-value"=p.value)
#Tidy up
rm(i,j,k,s,p,m,dfs,q,pair,diff,lwr,upr,q.stat,p.value)
```

Allowing for rounding and ordering, the contents of TK.Comps are the same as the results obtained using TukeyHSD.

	Pairs	Diff	Lower	Upper	q.Stat	p.value
1	mu4 - mu1	0.8237	0.1367	1.5107	4.7949	0.0113
2	mu5 - mu1	0.5954	-0.3077	1.4985	2.6366	0.3493
3	mu2 - mu1	0.5692	-0.1039	1.2423	3.3817	0.1343
4	mu3 - mu1	0.0000	-0.6731	0.6731	0.0000	1.0000
5	mu4 - mu3	0.8237	0.1367	1.5107	4.7949	0.0113
6	mu5 - mu3	0.5954	-0.3077	1.4985	2.6366	0.3493
7	mu2 - mu3	0.5692	-0.1039	1.2423	3.3817	0.1343
8	mu4 - mu2	0.2545	-0.4325	0.9415	1.4815	0.8319
9	mu5 - mu2	0.0262	-0.8769	0.9293	0.1160	1.0000
10	mu4 - mu5	0.2283	-0.6852	1.1418	0.9995	0.9540

Since
$$\mu_j - \mu_k = \tau_j - \tau_k,$$

this program (and the function `TukeyHSD`) works for the treatment effects model and yields identical results. Remember, `TukeyHSD` does not like the treatment means model.

On a side note, it is suggested by some that in order to avoid contradictory results in either of the Tukey procedures, a difference between a parameter pair is considered significant only if the closed interval for the estimated parameter difference in question does not lie in the closed interval formed by an estimated parameter difference for which the corresponding parameter pair has already been found not to be significantly different.

11.8.3   Scheffe's pairwise F-tests

Bounds for the confidence intervals and corresponding test statistics take on the appearance

$$\left(\hat{\mu}_j - \hat{\mu}_k\right) \pm s\,\sqrt{(p-1)\,F\left(\alpha; p-1, n-p\right)}\sqrt{\frac{1}{n_j} + \frac{1}{n_k}}$$

and

$$F_{jk}^* = \frac{\left(\hat{\mu}_j - \hat{\mu}_k\right)^2}{s^2(p-1)\left(\dfrac{1}{n_j} + \dfrac{1}{n_k}\right)},$$

with $df_N = p-1$ and $df_D = n-p$, respectively. The code for the Tukey–Kramer procedure can be altered to do the job.

Comments in the following code are limited to only those places where a change in the Tukey–Kramer code has been made.

```
mus<-with(OneWayData,tapply(y,Tr,mean))
ns<-with(OneWayData,tapply(y,Tr,length))
#Get number of samples and entry names — ordering not needed
p<-length(mus); labs<-c("mu1","mu2","mu3","mu4","mu5")
#Store this stuff and clean house a bit — no positions included
stuff<-data.frame(mus,ns,row.names=labs); rm(mus,ns)
k<-1; m<-choose(5,2)
s<-tc.sum$sigma
#Get degrees of freedom for F-distribution
df1<-p-1; df2<-tc.sum$df[2]
#Get the "F-critical value" for interval estimates
F.crit<-qf(.05, df1, df2, lower.tail = F)
```

```
pair<-c(rep(" ",m)); diff<-numeric(m);
lwr<-numeric(m); upr<-numeric(m)
F.stat<-numeric(m); p.value<-numeric(m)
attach(stuff)
#Begin outer loop, note change to counter
for (i in 1:(p-1)){
        #Begin inner loop, note change to counter
        for (j in (i+1):p){
                pair[k]<-paste(row.names(stuff[i,]),
                        "-",row.names(stuff[j,]))
                diff[k]<-round(mus[i]-mus[j],4)
                #Compute lower bound for F-interval
                lwr[k]<-round(diff[k]-
                        s*sqrt(F.crit*(p-1)*(1/ns[i]+1/ns[j])),4)
                #Compute upper bound for F-interval
                upr[k]<-round(diff[k]+
                        s*sqrt(F.crit*(p-1)*(1/ns[i]+1/ns[j])),4)
                #Get F-statistic
                F.stat[k]<-round(diff[k]^2/
                        (s^2*(p-1)*(1/ns[i]+1/ns[j])),4)
                #Get p-value using F-distribution
                p.value[k]<-round(pf(F.stat[k],
                        df1,df2,lower.tail=FALSE),4)
                k<-k+1}}
detach(stuff)
Scheffe.Comps<-data.frame("Pairs"=pair,"Diff"=diff,
  "Lower"=lwr,"Upper"=upr,"F-Stat"=F.stat,"p-value"=p.value)
```

The contents of `Scheffe.Comps` are

	Pairs	Diff	Lower	Upper	F.Stat	p.value
1	mu1 - mu2	-0.5692	-1.3299	0.1915	1.4295	0.2376
2	mu1 - mu3	0.0000	-0.7607	0.7607	0.0000	1.0000
3	mu1 - mu4	-0.8237	-1.6001	-0.0473	2.8738	0.0319
4	mu1 - mu5	-0.5954	-1.6160	0.4252	0.8690	0.4890
5	mu2 - mu3	0.5692	-0.1915	1.3299	1.4295	0.2376

```
 6  mu2 - mu4 -0.2545 -1.0309  0.5219 0.2743  0.8932
 7  mu2 - mu5 -0.0262 -1.0468  0.9944 0.0017  1.0000
 8  mu3 - mu4 -0.8237 -1.6001 -0.0473 2.8738  0.0319
 9  mu3 - mu5 -0.5954 -1.6160  0.4252 0.8690  0.4890
10  mu4 - mu5  0.2283 -0.8041  1.2607 0.1249  0.9728
```

11.8.4   Brown–Forsyth test

A small adjustment to the function `bf.Ftest` constructed in Chapter 6 produces a function that works for factorial models.

```
Anova.bf.Ftest <-function(e,group){
#Get test data and factor names
dname<-deparse(substitute(e))
xname<-deparse(substitute(group))
#Store test data and factor levels in a data frame
e.list<-data.frame(e,group)
#Compute group medians and deviations from medians
meds<-with(e.list,tapply(e,group,median))
dev<-with(e.list,abs(e-meds[group]))
#Fit to a linear model
info<-anova(lm(dev~group,e.list))
#Extract and store results
results<-list(statistic=c(F=info[1,4]),
  parameters=c(df=info[,1]),p.value=info[1,5],
   method=paste("Brown-Forsyth test across factor",
    xname),data.name=dname,alternative=c("Variances are
unequal"))
class(results)<-"htest";return(results)}
```

Then

```
> e<-residuals(tc.mod)
> with(OneWayData,Anova.bf.Ftest(e,Tr))
   Brown-Forsyth test across factor Tr
data: e
F = 1.3025, df1 = 4, df2 = 51, p-value = 0.2816
alternative hypothesis: Variances are unequal
```

duplicates the results from function hov.

11.8.5 Generating one-factor data to play with

Datasets of the form of `OneWayData` might be generated using code of the form shown below.

```
#First clear the workspace
rm(list=ls(all=T))
#Initialize response and factor variable
y<-NULL; Tr<-NULL
#Start for-loop
for (j in 1:5) {
    #Get a cell size
    k<-trunc(runif(1,1,6))
    #Identify factor level
    Tr<-c(Tr,rep(j,k))
    #Randomly assign a treatment mean
    mu<-runif(1,2,4)
    #Get "observed" responses
    y<-c(y,rnorm(k,mean=mu,sd=0.25))}
    #Repeat the above for j = 1 to 5
#Store y and Tr in a data frame
OneWayData<-data.frame(y,Tr)
#Define Tr as a factor
OneWayData$Tr<-factor(OneWayData$Tr)
#Take a look at the summary by treatments
with(OneWayData,tapply(y,Tr,summary))
```

The `dump` function can then be used to save the source code for the data frame, if so desired.

Chapter 12

One-Factor Models with Covariates

12.1 Introduction 283
12.2 Exploratory Data Analysis 284
12.3 Model Construction and Fit 286
12.4 Diagnostics 289
    12.4.1 The constant variance assumption 289
    12.4.2 The normality assumption 291
    12.4.3 The independence assumption 292
    12.4.4 The presence and influence of outliers 292
12.5 Pairwise Comparisons of Treatment Effects 294
    12.5.1 With a control treatment 294
    12.5.2 Without a control treatment 295
12.6 Models with Two or More Covariates 297
12.7 For the Curious 298
    12.7.1 Scheffe's pairwise comparisons 298
    12.7.2 The centered model 299
    12.7.3 Generating data to play with 300

12.1 Introduction

Let Y represent a response variable of interest that is to be subjected to p treatments within a categorical explanatory variable. Suppose, further, that it is known (or at least suspected) that the response also depends linearly on q continuous explanatory variables, $X_1, X_2, \ldots, X_q$, also referred to as *covariates*. For $j = 1, 2, \ldots, p$ and $i = 1, 2, \ldots, n_j$, denote the ith observed response within the jth treatment by y_{ij}, and denote the corresponding levels of the covariates by $x_{ij1}, x_{ij2}, \ldots, x_{ijq}$. As with one-factor models, although desireable, treatment sample sizes, n_j, need not be equal in size. Two equivalent models are presented.

In the default setting, R uses treatment coding as for the one-factor case. For this approach, the starting model has the form

$$y_{ij} = \beta_0 + \theta_j + \beta_1 x_{ij1} + \beta_2 x_{ij2} + \cdots + \beta_q x_{ijq} + \varepsilon_{ij},$$

and an equivalent full rank model is obtained by removing parameter redundancy through setting one of the θ_j equal to zero. The parameters β_k, $k \neq 0$, represent the regression coefficients for the covariates and the θ_j provide a means of getting at treatment effects. In the default setting, the first treatment parameter is removed, resulting in a model having the appearance

$$y_{ij} = \begin{cases} \beta_0 + \beta_1 x_{ij1} + \beta_2 x_{ij2} + \cdots + \beta_q x_{ijq} + \varepsilon_{ij} & \text{for } j = 1 \\ \beta_0 + \theta_j + \beta_1 x_{ij1} + \beta_2 x_{ij2} + \cdots + \beta_q x_{ijq} + \varepsilon_{ij} & \text{for } j \neq 1. \end{cases}$$

As with all models discussed thus far, the matrix form for this model is

$$\mathbf{y} = \mathbf{X}\boldsymbol{\beta} + \boldsymbol{\varepsilon}$$

with all terms being as defined earlier. Moreover, with the application of the treatment coding scheme, the design matrix acquires full column rank.

An alternative to the above approach is the *centered model*, which uses centered values of the covariates and permits the isolation and estimation of the overall mean response. In algebraic form, this model has the appearance

$$y_{ij} = \mu + \tau_j + \gamma_1 (x_{ij1} - \bar{x}_1) + \gamma_2 (x_{ij2} - \bar{x}_2) + \cdots + \gamma_q (x_{ijq} - \bar{x}_q) + \varepsilon_{ij},$$

where μ is the overall mean response parameter. The τ_j represent treatment effects parameters and the γ_k are regression parameters for the covariates.

The design matrix for this model does not have full column rank, but steps can be taken to remedy this. If the data are balanced (by treatments) the *sum contrast condition*, $\sum_{j=1}^{p} \tau_j = 0$, can be used to obtain an equivalent full rank model. If the data are unbalanced, by treatments, and if there is concern this might exercise influence on results, a *weighted sum contrast condition* of the form $\sum_{j=1}^{p} w_j \tau_j = 0$ *may* be used. However, the literature seems to suggest that this is typically not done.

In either case, as for earlier models, for all $j = 1, 2, \ldots, p$ and $i = 1, 2, \ldots, n_j$, it is assumed that error terms, ε_{ij}, are independent, and are identically and normally distributed with $\varepsilon_{ij} \sim N(0, \sigma^2)$. For covariance models, there is also the added assumption that the response variable Y is (at least approximately) linearly related to each of the covariates $X_1, X_2, \ldots, X_q$.

The majority of the discussion in this chapter considers a one-factor experiment design having a single covariate. Methods for one-factor experiments having more than one covariate are analogous.

If the variable for the factor is denoted by Tr, data for such an experiment might have the appearance of the data shown in Table 12.1. These data are stored in the file `Data12x01.R` as a data frame named `AncovaData`.

12.2 Exploratory Data Analysis

As with earlier models, summary statistics are easily obtained using commands of the form

```
summary(AncovaData)
with(AncovaData,by(AncovaData,Tr,summary))
```

This is also a good point at which to see if numerically coded treatments, such as 1, 2, 3, etc., are defined as factors.

TABLE 12.1: Data for Chapter 12 Illustrations

Tr1		Tr2		Tr3	
Y	X	Y	X	Y	X
24.48	14.4	20.24	8.7	29.80	14.5
21.90	12.0	16.46	5.3	35.10	19.0
25.50	15.0	28.82	15.6	27.70	12.5
31.34	19.7	32.62	18.6	27.12	12.1
21.64	11.7	28.12	15.1	23.88	9.4
20.82	11.1	18.96	7.3	29.68	13.9
23.68	13.4	32.12	18.6	34.32	18.1
19.38	9.9	25.72	13.1	31.82	16.1
19.44	9.7	18.54	7.2	34.72	18.6
27.82	17.1	22.74	10.7	31.32	15.6

It is useful to check if the data suggest an approximate linear relationship between the response variable and each of the covariates (see Figure 12.1).

```
plot(y~x,AncovaData,pch=unclass(Tr),main=NULL)
legend(5,35,legend=c("Tr1","Tr2","Tr3"),
                    pch=c(1,2,3),bty="n")
```

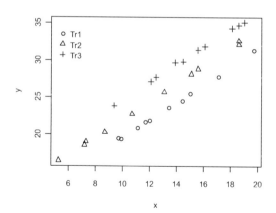

FIGURE 12.1: Scatterplot of dataset `AncovaData` with plotted points identified by treatments.

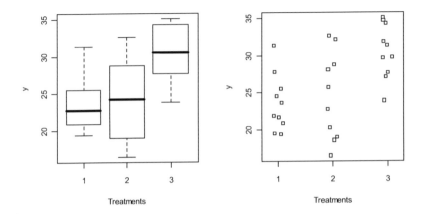

FIGURE 12.2: Boxplots and a jittered stripchart of the response variable by treatments.

Two other plots that can be useful are of the responses by treatments (see Figure 12.2) using the code

```
with(AncovaData,plot(Tr,y,xlab="Treatments",ylab="y"))
with(AncovaData,stripchart(y~Tr,vertical=T,
        method="jitter",xlab="Treatments",ylab="y"))
```

The purposes of these plots are as mentioned for one-factor models in the previous chapter.

Fitting, testing, and estimation for the default (uncentered) and centered models are now looked at seperately.

12.3   Model Construction and Fit

For the default setting, the data are fitted to a model of the form

$$y_{ij} = \beta_0 + \theta_j + \beta_1 x_{ij} + \varepsilon_{ij}$$

and treatment coding is used to remove the redundancy in the treatment parameters. In the default setting, θ_1 is removed (i.e., $\theta_1 = 0$). Thus, β_0 represents the y-intercept for this model when $j = 1$, β_1 represents the regression parameter for the covariate (or slope of the corresponding regression lines), and θ_j (for $j \neq 1$) represents the equivalent of a vertical translation which serves as a measure of the difference between the effects of treatments $j \neq 1$ and treatment 1.[1]

The model fitting process is as before; the commands

```
ancova.mod<-lm(y~Tr+x,AncovaData)

ancova.sum<-summary(ancova.mod)
```

produce all information needed to perform the relevant analyses. The coefficients in `ancova.mod` show

```
> ancova.mod$coefficients

(Intercept)         Tr2         Tr3            x
   7.622268    2.479468    5.062058    1.192368
```

which provides the fitted model

$$\hat{y} = \begin{cases} 7.622 + 1.19x & \text{for } j = 1 \\ 10.102 + 1.19x & \text{for } j = 2 \\ 12.684 + 1.19x & \text{for } j = 3 \end{cases}$$

There are some additional tasks when dealing with covariance models.

Denote parameters for interaction effects between levels of the factor and the covariate by $(\theta\beta)_{j1}$. Then the *constant slope assumption* can be tested via a partial F-test on the hypotheses

$$H_0 : y_{ij} = \beta_0 + \theta_j + \beta_1 x_{ij} + \varepsilon_{ij} \quad \text{vs.}$$
$$H_1 : y_{ij} = \beta_0 + \theta_j + \beta_1 x_{ij} + (\theta\beta)_{j1} x_{ij} + \varepsilon_{ij}$$

Thus,

```
> int.model<-lm(y~Tr+x+x:Tr,AncovaData)

> anova(ancova.mod,int.model)

Analysis of Variance Table

Model 1: y ~Tr + x

Model 2: y ~Tr + x + x:Tr
```

	Res.Df	RSS	Df	Sum of Sq	F	Pr(>F)
1	26	0.84304				
2	24	0.77306	2	0.069981	1.0863	0.3535

---

[1] If necessary, reset the contrast and treatment coding settings for the function `lm` to the default settings using the functions `contr.treatment`, `C`, and `options` as described in Section 11.3. Another way is to simply restart R.

The last line of the output, corresponding to model 2, indicates the contribution significance of the interaction term. Here the constant slope assumption is not rejected.

Note that the contribution significance of the covariate can also be tested using an analogous partial F-test.

The significance of the model, or the hypotheses

$$H_0 : y_{ij} = \beta_0 + \varepsilon_{ij} \quad \text{vs.}$$
$$H_1 : y_{ij} = \beta_0 + \theta_j + \beta_1 x_{ij} + \varepsilon_{ij},$$

can also be tested using a partial F-test; however, the same result can be obtained simply by looking at the last line in ancova.sum,

```
F-statistic: 8718 on 3 and 26 DF, p-value: < 2.2e-16
```

To assess the significance of treatment effects, the hypotheses

$$H_0 : y_{ij} = \beta_0 + \beta_1 x_{ij} + \varepsilon_{ij} \quad \text{vs.}$$
$$H_1 : y_{ij} = \beta_0 + \theta_j + \beta_1 x_{ij} + \varepsilon_{ij}$$

can be tested using

```
> reg.model<-lm(y~x,AncovaData)
> anova(reg.model,ancova.mod)
Analysis of Variance Table
Model 1: y ~x
Model 2: y ~Tr + x
  Res.Df      RSS  Df  Sum of Sq       F      Pr(>F)
1     28  125.194
2     26    0.843   2     124.35  1917.5  < 2.2e-16 ***
```

The last line of the output, corresponding to model 2, indicates treatment effects can be considered significant.

In package HH the routine ancova, an lm wrapper for one-factor analysis of covariance, is available. This function also produces a plot of the fitted model. The basic call for this function is

```
ancova(y~Tr*x, dataset )
```

if the constant slope assumption needs to be tested, and

```
ancova(y~x+Tr, dataset )
```

if the significance of treatment effects is to be tested on the constant slope model.

Don't forget to `detach` package `HH` and its accompanying packages in the correct order when done with `ancova`, just in case conflicts arise with later loaded packages.

Caution: There is one point to note about the function `ancova`; always place the term of interest last in the model formula to ensure appropriate results.[2]

12.4 Diagnostics

Figure 12.1 serves as a first diagnostic graph to check if the observed responses exhibit an approximate linear relationship with the covariate. The remaining diagnostics, in the case of a single covariate, include a combination of methods used for simple regression and one-factor models.

Begin by gathering the statistics to be used,

```
y.hat<-fitted.values(ancova.mod)

e<-residuals(ancova.mod)

h<-hatvalues(ancova.mod)

r<-e/(ancova.sum$sigma*sqrt(1-h))

d<-rstudent(ancova.mod)
```

Keep in mind that it is very possible not all of the above will be needed. Also, many prefer to use studentized residuals in the following analyses. Numerical backup tests can be performed using the earlier discussed tests for constant variances across treatments and by levels of the covariate if desired. Similarly, the tests of normality and methods for influence analysis also apply. However, for the most part, it appears that graphical methods are relied upon for covariance models. For those interested, a brief coverage of numerical tests and methods are also given here.

12.4.1 The constant variance assumption

Methods used to assess the validity of the constant variance assumption in linear regression and one-way ANOVA also apply here. Figure 12.3 was obtained using

---

[2]Compare, for example, the results obtained using partial F-tests against output obtained from `ancova(y~x+Tr,AncovaData)` and `ancova(y~Tr+x,AncovaData)`.

```
attach(AncovaData)
plot(e~y.hat,xlab="Fitted values",ylab="Residuals")
plot(e~x,xlab="x",ylab="Residuals")
stripchart(e~Tr,vertical=T,method="jitter",
                xlab="Treatment",ylab="Residuals")
detach(AncovaData)
```

Plotted points in the two left-hand plots can be represented by treatments by including the argument

```
pch=as.character(Tr)
```

in each `plot` function. While identifying plotted points by treatments can add more flavor to the plots, in this case the results are distracting; try it.

In situations when some points in the plots overlap, particularly for the plot containing fitted values, a plotting command of the form

```
plot(e~jitter(y.hat))
```

might help in interpreting the plot. Recall that the plot of the residuals against X and treatments can be useful for detecting other issues involving X and treatments. For example, in Figure 12.3 it can be observed that treatments 2 and 3 appear to have a couple of lonely cases.

The earlier constructed function, `bf.Ftest`, for the Brown–Forsyth test can be used to test the constant variances assumption across levels of the covariate, and the function `Anova.bf.Ftest` can be used to test across treatments. The function `hov` can also be used to test across treatments:

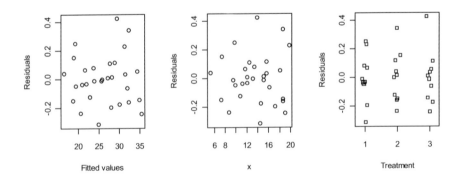

FIGURE 12.3: Scatterplots of residuals obtained from the fitted model `ancova.mod` against the fitted values, x, and factor levels.

```
> hov(e~Tr,AncovaData)

          hov: Brown-Forsyth

data: e

F = 0.0164, df:Tr = 2, df:Residuals = 27, p-value = 0.9837

alternative hypothesis: variances are not identical
```

and across levels of X as described in Chapter 6. Observe that executing

```
with(AncovaData,Anova.bf.Ftest(e,Tr))
```

produces the the same results.

12.4.2   The normality assumption

Procedures used to assess the validity of the normality assumption remain the same; see Figure 12.4, which was obtained using

```
qqnorm(r,main=NULL); abline(a=0,b=1,lty=3)
```

Recall that a `qqline` may be used as a reference line in place of the line $y = x$. The correlation coefficient test for normality can be performed or the Shapiro–Wilk normality test can be used. For example,

```
> qq.cortest(r,0.05)

      QQ Normal Probability Corr. Coeff. Test, alpha = 0.05

data: r, n = 30

RQ = 0.9855, RCrit = 0.964

alternative hypothesis: If RCrit > RQ, Normality assumption
is invalid
```

and

```
> shapiro.test(r)

      Shapiro-Wilk normality test

data: r

W = 0.97, p-value = 0.5394
```

Keep in mind that the function `qq.cortest` will have to be sourced before it can be used.

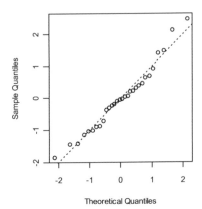

FIGURE 12.4: QQ normal probability plot of the studentized residuals for the model `ancova.mod`.

12.4.3   The independence assumption

Time dependence and the absence of other important explanatory variables (more covariates), which might result in an apparent violation of the independence assumption, can also be assessed *if* information on time and other variables in relation to the observed data are available. Just as with earlier models, plots of $(\hat{\varepsilon}_i, \hat{\varepsilon}_{i+1})$ pairs as well as plots of the residuals against time and other variables serve to identify possible violations of the independence assumption.

In the case of time-series data, suggestions applicable to simple regression models and one-factor models apply here, too. The Durbin–Watson test may be used if applicable.

12.4.4   The presence and influence of outliers

Unlike one-factor models, the presence of the covariate means outlying values of the covariate should be looked at as well in the search for possible influential outliers. Plots of the form of Figure 12.1 provide a preliminary step in flagging cases that appear to go against the general trend. A plot of the studentized deleted residuals, such as Figure 12.5, helps with identifying outlying observed responses that might be influential. The Bonferroni cutoff lines included in the plot suggest that of the three (10%) extreme cases, Case 26 might be of concern. Code to obtain Figure 12.5 is shown below.

```
plot(d~y.hat,xlab="Fitted values",
     ylab="Studentized deleted residuals",ylim=c(-3,3))
```

```
identify(y.hat,d)

m<-ceiling(.1*length(d));df<-length(d)-4-1

a<-0.05/m;  cv<-qt(a/2,df,lower.tail=F)

abline(h=c(-cv,0,cv),lty=c(3,1,3))
```

The last three lines provide Bonferroni cutoff lines. Even though Case 26 is flagged as an outlier, notice that Figure 12.1 suggests there are no cases that really stand out as being influential. If so desired, the earlier encountered function `influence.measures` can be used to further examing Case 26. For example, executing the commands

```
inf.tests<-data.frame(influence.measures(ancova.mod)[2])

inf.tests[26,]
```

provides influence measures (not shown here) only for Case 26.

As with simple regression and one-factor models, graphical methods are often quite adequate for determining the potential influence of an outlying case. Plots such as Figures 12.1 and 12.5 typically prove more than adequate. Note that half-normal plots can also be used in identifying potential outliers.

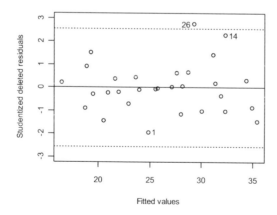

FIGURE 12.5: Plot of studentized deleted residuals of the fitted model `ancova.mod`. The three (10%) most extreme cases are labeled and tested using Bonferroni cutoff values.

12.5　Pairwise Comparisons of Treatment Effects

What this process essentially boils down to is testing the differences between y-intercepts of the treatment specific regression models (see Figure 12.6). A drawback here is that the earlier `pairwise.t.test` function cannot be used since it does not account for effects contributed by the covariate.

Here is some groundwork before considering the two possible scenarios as described for one-factor designs. Since the default setting uses $\theta_1 = 0$ in

$$y_{ij} = \beta_0 + \theta_j + \beta_1 x_{ij} + \varepsilon_{ij},$$

the piecewise definition of the model under consideration by factor levels has the following appearance:

$$y_{ij} = \begin{cases} \beta_0 + \beta_1 x_{ij} + \varepsilon_{ij} & \text{for } j = 1, \\ \beta_0 + \theta_2 + \beta_1 x_{ij} + \varepsilon_{ij} & \text{for } j = 2, \\ \beta_0 + \theta_3 + \beta_1 x_{ij} + \varepsilon_{ij} & \text{for } j = 3. \end{cases}$$

Comparing with the corresponding unadjusted centered (effects) model,

$$y_{ij} = \mu + \tau_j + \gamma (x_{ij} - \bar{x}) + \varepsilon_{ij},$$

it can be shown that parameters for the two models satisfy

$$\beta_1 = \gamma, \ \beta_0 = \mu + \tau_1 - \gamma \bar{x},$$
$$\beta_0 + \theta_2 = \mu + \tau_2 - \gamma \bar{x}, \text{ and } \beta_0 + \theta_3 = \mu + \tau_3 - \gamma \bar{x}.$$

These lead to

$$\tau_2 - \tau_1 = \theta_2, \ \tau_3 - \tau_1 = \theta_3 \text{ and } \tau_2 - \tau_3 = \theta_2 - \theta_3$$

which are useful in constructing code for pairwise tests. Begin with the simpler of the two scenarios.

12.5.1　With a control treatment

If treatment 1 is the control treatment, then all other treatments are compared against this and not with each other. For the given data, the above discussion indicates that an adequate (simultaneous) assessment of such pairwise comparisons is to test the hypotheses, for $j = 2$ and 3,

$$H_0 : \theta_j = 0 \quad \text{vs.} \quad H_1 : \theta_j \neq 0.$$

Results for this test are given in

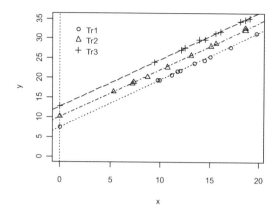

FIGURE 12.6: Plot of dataset `AncovaData` along with plots of the lines representing the fitted model by treatments. Points plotted on the line $x = 0$ represent intercepts.

```
> ancova.sum$coefficients
```

	Estimate	Std. Error	t value	Pr(>|t|)
(Intercept)	7.622268	0.13421864	56.78994	8.514548e-29
Tr2	2.479468	0.08149613	30.42436	7.419188e-22
Tr3	5.062058	0.08179439	61.88760	9.255603e-30
x	1.192368	0.00907021	131.45980	3.077020e-38

The lines for `Tr2` and `Tr3` provide the necessary information for the tests.

Note that nothing is said about the difference between treatments 2 and 3, so a quantitative ranking would be needed if one were interested in finding which of these two treatments is "better."

12.5.2 Without a control treatment

The three comparisons are covered by the relationships $\tau_2 - \tau_1 = \theta_2$, $\tau_3 - \tau_1 = \theta_3$, and $\tau_2 - \tau_3 = \theta_2 - \theta_3$, and so the equivalent null hypotheses in question are

$$H_0 : \theta_2 = 0; \quad H_0 : \theta_3 = 0; \quad \text{and} \quad H_0 : \theta_2 - \theta_3 = 0.$$

Bonferroni's procedure can be performed for all pairwise comparisons quite easily by making some simple alterations to earlier (Bonferroni) R code by substituting

$$\hat{\tau}_2 - \hat{\tau}_1 = \hat{\theta}_2, \quad \hat{\tau}_3 - \hat{\tau}_1 = \hat{\theta}_3, \quad \hat{\tau}_2 - \hat{\tau}_3 = \hat{\theta}_2 - \hat{\theta}_3,$$

and

$$s_{\hat{\tau}_2 - \hat{\tau}_1}^2 = s_{\hat{\theta}_2}^2, \qquad s_{\hat{\tau}_3 - \hat{\tau}_1}^2 = s_{\hat{\theta}_3}^2, \qquad s_{\hat{\tau}_2 - \hat{\tau}_3}^2 = s_{\hat{\theta}_2 - \hat{\theta}_3}^2 = s_{\hat{\theta}_2}^2 + s_{\hat{\theta}_3}^2 - 2s_{\hat{\theta}_2 \hat{\theta}_3}$$

in the intervals

$$(\hat{\tau}_j - \hat{\tau}_k) - t(\alpha/(2m), n - p - 1)\, s_{\hat{\tau}_j - \hat{\tau}_k}$$
$$< (\tau_j - \tau_k) <$$
$$(\hat{\tau}_j - \hat{\tau}_k) + t(\alpha/(2m), n - p - 1)\, s_{\hat{\tau}_j - \hat{\tau}_k}.$$

and the test statistics

$$T_{jk} = \left| \frac{\hat{\tau}_j - \hat{\tau}_k}{s_{\hat{\tau}_j - \hat{\tau}_k}} \right|.$$

Here is sample code to perform this task.

```
#Obtain pairs and their differences
Diff<-numeric(3); Pair<-c(rep(" ",3))
attach(ancova.mod)
Diff[1]<- coefficients[2];Pair[1]<-c("Tr2 - Tr1")
Diff[2]<- coefficients[3];Pair[2]<-c("Tr3 - Tr1")
Diff[3]<-coefficients[2]-coefficients[3]
Pair[3]<-c("Tr2 - Tr3")
detach(ancova.mod)
#Obtain variances and covariance
attach(ancova.sum)
var.Theta2<-sigma^2*cov.unscaled[2,2]
var.Theta3<-sigma^2*cov.unscaled[3,3]
cov.Theta23<-sigma^2*cov.unscaled[2,3]
detach(ancova.sum)
#Obtain standard errors and critical value
SE<-numeric(3)
SE[1]<-sqrt(var.Theta2);SE[2]<-sqrt(var.Theta3)
SE[3]<-sqrt(var.Theta2+var.Theta3-2*cov.Theta23)
crit.val<-qt(.05/6,26,lower.tail=F)
#Construct interval bounds, test values, and p-values
LB<-numeric(3); UB<-numeric(3)
TV<-numeric(3); PV<-numeric(3)
for (i in (1:3)){
```

```
LB[i]<-Diff[i]-crit.val*SE[i]
UB[i]<-Diff[i]+crit.val*SE[i]
TV[i]<-abs(Diff[i]/SE[i])
PV[i]<-2*pt(TV[i],26,lower.tail=F)}
```
#Round down results for readability
```
for (i in (1:3)){
    Diff[i]<-round(Diff[i],3); SE[i]<-round(SE[i],3)
    LB[i]<-round(LB[i],3);UB[i]<-round(UB[i],3)
    TV[i]<-round(TV[i],3); PV[i]<-round(PV[i],3)}
```
#Store results
```
results<-data.frame(cbind(Pair,Diff,SE,LB,UB,TV,PV))
```

Then

```
> results
        Pair    Diff    SE     LB     UB      TV PV
1 Tr2 - Tr1   2.479 0.081  2.271  2.688 30.424  0
2 Tr3 - Tr1   5.062 0.082  4.853  5.271 61.888  0
3 Tr2 - Tr3  -2.583 0.085   -2.8 -2.365 30.424  0
```

In the case where the number of pairwise comparisons is large, the Bonferroni procedure might be replaced by Scheffe's procedure. Sample code for Scheffe's procedure is shown in the last section of this chapter. Tukey's procedure is not appropriate for covariance models.

12.6 Models with Two or More Covariates

The inclusion of two or more covariates adds more layers to the analysis, but the basic methods and coding remain essentially the same as for multiple regression and one-factor models. The lm function call for the uncentered nonconstant slope model has the form

```
lm(y~Tr + x1 + x2 +···+ xq
       + x1:Tr + x2:Tr +···+ xq:Tr, data.set)
```

and for the constant slope model

```
lm(y~Tr + x1 + x2 +···+ xq, data.set)
```

Reduced models for partial F-tests are obtained by simply excluding the variable of interest from `ancova.mod`.

The issues of influential outliers and multicollinearity probably play a larger role as the number of covariates in a model increases. In such cases, the methods from multiple regression work well.

12.7 For the Curious

If the number of treatments is large, Scheffe's procedure may be preferred over Bonferroni's procedure when performing simultaneous pairwise comparisons.

12.7.1 Scheffe's pairwise comparisons

Scheffe's procedure makes use of the intervals

$$(\hat{\tau}_j - \hat{\tau}_k) - \sqrt{(p-1)\,F(\alpha, p-1, n-p-1)}\, s_{\hat{\tau}_j - \hat{\tau}_k}$$
$$< (\tau_j - \tau_k) <$$
$$(\hat{\tau}_j - \hat{\tau}_k) + \sqrt{(p-1)\,F(\alpha, p-1, n-p-1)}\, s_{\hat{\tau}_j - \hat{\tau}_k}$$

and the test statistics

$$T_{jk} = \frac{(\hat{\tau}_j - \hat{\tau}_k)^2}{(p-1)\, s^2_{\hat{\tau}_j - \hat{\tau}_k}}.$$

So, the only changes needed in the earlier code for Bonferroni's procedure are where $t(\alpha/(2m), n-p-1)$, T_{jk}, and the p-values are calculated. The adjusted commands are

```
crit.val<-sqrt(2*qf(.05,2,26,lower.tail=F))
TV[i]<-Diff[i]^2/(2*SE[i]^2)
PV[i]<-pf(TV[i],2,26,lower.tail=F)
```

Making these changes and then running the code produces

	Pair		Diff	SE	LB	UB	TV	PV
1	Tr2 -	Tr1	2.479	0.081	2.268	2.691	462.821	0
2	Tr3 -	Tr1	5.062	0.082	4.85	5.274	1915.038	0
3	Tr2 -	Tr3	-2.583	0.085	-2.803	-2.362	462.809	0

12.7.2 The centered model

Except for an initial difference in how R is instructed to fit the data to the model, much remains the same as for the uncentered model. For the current data, the unadjusted centered model has the appearance

$$y_{ij} = \mu + \tau_j + \gamma\left(x_{ij} - \bar{x}\right) + \varepsilon_{ij},$$

with the sum contrast condition $\sum_{j=1}^{3} \tau_j = 0$. To fit the model, R is first alerted to use the sum contrast by the `options` function

```
options(contrasts=c("contr.sum","contr.poly"))
ancova.mod<-lm(y~Tr+I(x-mean(x)),AncovaData)
```

When using the `contr.sum` option, the R default setting removes the last treatment. The various other models for purposes of partial F-tests are obtained using analogous commands:

```
int.model<-lm(y~Tr+I(x-mean(x))+
                Tr:I(x-mean(x)),AncovaData)
red.model<-lm(y~1,AncovaData)
reg.model<-lm(y~I(x-mean(x)),AncovaData)
```

The diagnostics remain identical since the various statistics remain the same.

Pairwise comparisons of effects parameters in the centered model may be performed best by comparing differences between mean responses at $x = \bar{x}$ since, for this value of the covariate, the mean response for each treatment is given by $\mu + \tau_j$; that is, at $x = \bar{x}$ the model reduces to

$$y_{ij} = \mu + \tau_j + \varepsilon_{ij}.$$

The necessary statistics for computing interval estimates and performing pairwise comparisons are also obtained from the `lm` and model `summary` objects. The comparisons are as before,

$$\text{H}_0 : \tau_1 - \tau_2 = 0, \quad \text{H}_0 : \tau_1 - \tau_3 = 0, \quad \text{and} \quad \text{H}_0 : \tau_2 - \tau_3 = 0.$$

However, the sum contrast condition requires that $\tau_3 = -\tau_1 - \tau_2$. This is then incorporated into the computations of

$$s_{\hat{\tau}_1 - \hat{\tau}_2}^2, \qquad s_{\hat{\tau}_1 - \hat{\tau}_3}^2 = s_{2\hat{\tau}_1 + \hat{\tau}_2}^2, \qquad s_{\hat{\tau}_2 - \hat{\tau}_3}^2 = s_{\hat{\tau}_1 + 2\hat{\tau}_2}^2$$

for use in

$$(\hat{\tau}_j - \hat{\tau}_k) - t(\alpha/(2m), n - p - 1)\, s_{\hat{\tau}_j - \hat{\tau}_k}$$
$$< (\tau_j - \tau_k) <$$
$$(\hat{\tau}_j - \hat{\tau}_j) + t(\alpha/(2m), n - p - 1)\, s_{\hat{\tau}_j - \hat{\tau}_k}.$$

and

$$T_{jk} = \left| \frac{\hat{\tau}_j - \hat{\tau}_k}{s_{\hat{\tau}_j - \hat{\tau}_k}} \right|.$$

With these adjustments, the code for the uncentered case does the job.

Code may similarly be written to use Scheffe's method to look at differences between treatment effects at $x = \bar{x}$.

12.7.3   Generating data to play with

The data used in this chapter were generated using the following code.

```
#Initialize variables and "observation" counter
Tr<-integer(30); y<-numeric(30); x<-numeric(30)
k<-1
#Start outer data generation loop for treatments
for (j in 1:3){
    #Get some x-values
    xx<-round(runif(10,5,20),1)
    #Assign a mu (can be made random)
    mu<-3.5*j
    #Get random errors
    e<-round(rnorm(10,mean=mu,sd=.5),1)
    #Start inner loop within treatments
    for (i in 1:10){
        Tr[k]<-j
        y[k]<-5+1.2*xx[i]+e[i]
        x[k]<-xx[i]
        k<-k+1}}
#Store results
AncovaData<-data.frame(y,x,Tr)
#Define Tr as a factor
AncovaData$Tr<-factor(AncovaData$Tr)
#Plot to see how the data look
plot(y~x,AncovaData,pch=unclass(Tr),main=NULL)
```

Again, the source code for the data frame can be saved if so desired.

Chapter 13

One-Factor Models with a Blocking Variable

13.1 Introduction 301
13.2 Exploratory Data Analysis 303
13.3 Model Construction and Fit 305
13.4 Diagnostics 306
    13.4.1 The constant variance assumption 307
    13.4.2 The normality assumption 308
    13.4.3 The independence assumption 309
    13.4.4 The presence and influence of outliers 309
13.5 Pairwise Comparisons of Treatment Effects 310
    13.5.1 With a control treatment 310
    13.5.2 Without a control treatment 311
13.6 Tukey's Nonadditivity Test 312
13.7 For the Curious 314
    13.7.1 Bonferroni's pairwise comparisons 314
    13.7.2 Generating data to play with 316

13.1 Introduction

For the following it is assumed that all cells contain at least one entry, so only *randomized complete block designs* are considered, as opposed to *incomplete block designs*.

Consider an experiment having a factor of interest, say factor A, and a *blocking variable*, say B. Suppose A has I levels and B has J levels. Let Y represent a response variable of interest and y_{ijk} the kth randomly observed response value for the ith level of factor A and j^{th} level of B, or the kth observed response in the ij^{th} cell. Use n_{ij} to represent the total number of observations within the ij^{th} cell and $n = \sum_{i=1}^{I} \sum_{j=1}^{J} n_{ij}$ the total number of observed responses.

Begin with the *cell means model*, a natural extension of the treatment means model for one-factor designs. Let each cell mean be denoted by the parameter μ_{ij}, then for $i = 1, 2, \ldots, I$, $j = 1, 2, \ldots, J$, and $k = 1, 2, \ldots, n_{ij}$ each observed response y_{ijk} may be expressed as the sum of the corresponding cell mean plus a random error term ε_{ijk},

$$y_{ijk} = \mu_{ij} + \varepsilon_{ijk}.$$

Just as the cell means model is a natural extension of the treatment means model from one-factor designs, a natural extension of the treatment effects model in one-factor designs exists for the block design in question.

For the design under consideration, effects on the response variable can be viewed as having two main sources: those due to the levels in factor A and those due to the levels in the blocking variable B. These effects are referred to as *main* and *blocking effects*, respectively. Thus, a main (or blocking) effect is defined to be the change produced in the response Y as a result of a change in factor levels (or block levels) *within the factor (block)*. Another source of effects, *interaction effects*, exists if the levels of the factor and the levels of the blocking variable interact in a manner that influences the observed response. The cell means model can be altered to yield a model that isolates the main, blocking, and interaction effects in an experiment.

For $i = 1, 2, \ldots, I$ and $j = 1, 2, \ldots, J$, let τ_{A_i} represent the main effects resulting from the ith level of A, τ_{B_j} the blocking effects resulting from the jth level of B and let $\tau_{AB_{ij}}$ represent the interaction effects (if any) of the ith level of A and the jth level of B. Then each cell mean may be viewed as the overall mean plus main, blocking, and interaction effects parameters,

$$\mu_{ij} = \mu + \tau_{A_i} + \tau_{B_j} + \tau_{AB_{ij}}.$$

Substitution of this expression in the cell means model yields the *nonadditive effects model*,

$$y_{ijk} = \mu + \tau_{A_i} + \tau_{B_j} + \tau_{AB_{ij}} + \varepsilon_{ijk},$$

where the parameters μ, τ_{A_i}, τ_{B_j}, and $\tau_{AB_{ij}}$ are all unknown constants. If $\tau_{AB_{ij}} = 0$, the result is the preferred *additive effects model*,

$$y_{ijk} = \mu + \tau_{A_i} + \tau_{B_j} + \varepsilon_{ijk}.$$

While all three models described above have a matrix representation

$$\mathbf{y} = \mathbf{X}\boldsymbol{\beta} + \boldsymbol{\varepsilon},$$

with terms being as previously defined, only the cell means model has a design matrix with full column rank. Thus, conditions need to be imposed to remove dependent columns in the design matrix (or redundant parameters in the model). As with the one-factor setting, there are equivalent ways in which this can be accomplished. Because of the considerable increase in complexity, the only approach presented here is the default treatment coded setting used by R.[1] The approach is analogous to that used for the one-factor treatment effects model.

For $i = 1, 2, \ldots, I$, $j = 1, 2, \ldots, J$, and $k = 1, 2, \ldots, n_{ij}$, the *nonadditive treatment coded model* has the appearance

$$y_{ijk} = \beta_0 + \beta_{A_i} + \beta_{B_j} + \beta_{AB_{ij}} + \varepsilon_{ijk},$$

---

[1] Be sure to reset the contrast to the default setting to duplicate results in this chapter. An easy way to do this is to simply restart R.

subject to the simplifications $\beta_{A_1} = 0$, $\beta_{B_1} = 0$, and $\beta_{AB_{1j}} = \beta_{AB_{i1}} = 0$. It should be remembered that the resulting parameters are not the same as the earlier clearly defined parameters μ, τ_{A_i}, τ_{B_j}, and $\tau_{AB_{ij}}$. The *additive treatment coded model* has the appearance

$$y_{ijk} = \beta_0 + \beta_{A_i} + \beta_{B_j} + \varepsilon_{ijk}$$

subject to the simplifications $\beta_{A_1} = \beta_{B_1} = 0$.

Assumptions for any of the above models are analogous to those of previous models; that is, the error terms are independent, and are identically and normally distributed with $\varepsilon_{ijk} \sim N(0, \sigma^2)$. Additionally, it is assumed that interactions between factor and block levels are absent (or negligible at worst).

There are two scenarios with respect to data. The default coding scheme applies to both scenarios. Consider first the scenario when at least one cell contains more than one observation. Since the continuous response variable, Y, is viewed as being dependent on the two categorical variables A and B, such data might be presented in the format shown in Table 13.1.

TABLE 13.1: Dataset 1 for Chapter 13 Illustrations.

B	1			2			3			4		
A	1	2	3	1	2	3	1	2	3	1	2	3
	25	22	19	17	11	12	22	18	16	15	15	10
Y	20	21		18	19	15	20	20	17	17	11	
	23	18				12						

The data are stored in `Data13x01.R` as a data frame named `BlockData`. The variable A is assigned values 1, 2, and 3, and values for the variable B are 1, 2, 3, and 4. As mentioned earlier, in such cases it is a good idea to make sure these variables are defined as factors.

13.2 Exploratory Data Analysis

The presence of interaction between factor and block levels can be checked graphically using what are referred to as *interaction plots*. The R function used to obtain such plots, with basic arguments, is

```
interaction.plot(x.factor, trace.factor, response)
```

The argument *x.factor* represents the factor whose levels form the horizontal axis, *trace.factor* is the factor whose levels form the *traces* (or plotted

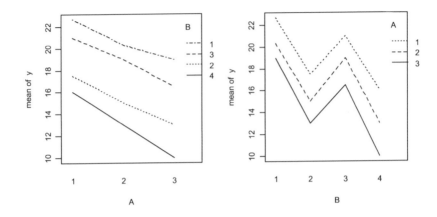

FIGURE 13.1: Interaction plots for dataset `BlockData` to determine whether interaction between the levels of the factor and blocking variable exist.

"lines"), and *response* represents the response variable, the mean of which forms the vertical axis.

It is useful to obtain two interaction plots, trading between the factor and the blocking variable for the *x.factor*. The commands[2]

```
with(BlockData,interaction.plot(A,B,y))
with(BlockData,interaction.plot(B,A,y))
```

produce Figure 13.1. The simplest interpretation of these plots is as follows. Since the traces in both of the interaction plots appear to be approximate vertical translations of each other, it is very likely that interaction effects are not significant. Moreover, the presence of distinct traces in both figures suggest that both treatment and blocking effects are probably significant. Be sure to refer to the relevant literature for more detailed discussions behind interpreting interaction plots.

Summaries of the data can be obtained as for one-factor designs using commands of the form:

```
summary(BlockData)
with(BlockData,tapply(y,A,summary))
by(BlockData,BlockData$B,summary)
```

The same goes for exploratory plots. Figure 13.2 provides a quick look at how the data behave within factor and blocking levels. These plots might be obtained using commands of the following form:

---

[2] Note that the functions `win.graph` and `par` can be used to format the window prior to plotting.

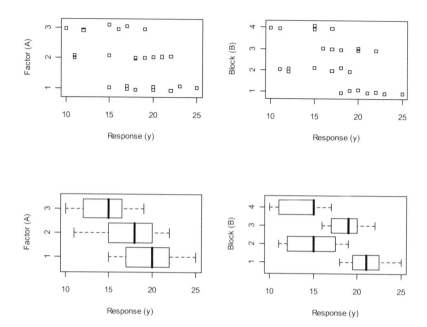

FIGURE 13.2: Stripcharts and boxplots of the response data by factor and blocking levels. Stripcharts are "jittered" to highlight repeated values of the response.

```
stripchart(y~A,method="jitter",xlab="Response (y)",
                              ylab="Factor (A)")
stripchart(y~B,method="jitter",xlab="Response (y)",
                              ylab="Block (B)")
plot(y~A,ylab="Response (y)",xlab="Factor (A)",
                              horizontal=T)
plot(y~B,ylab="Response (y)",xlab="Block (B)",horizontal=T)
```

13.3   Model Construction and Fit

If so desired (or required), a first step in the process might be to determine whether the interaction effects parameters are significant. With the *nonadditive treatment coded model*, this involves testing

$$H_0 : y_{ijk} = \beta_0 + \beta_{A_i} + \beta_{B_j} + \varepsilon_{ijk} \quad \text{vs.}$$

$$H_1 : y_{ijk} = \beta_0 + \beta_{A_i} + \beta_{B_j} + \beta_{AB_{ij}} + \varepsilon_{ijk}.$$

This task is accomplished simply enough by means of a partial F-test or by executing

```
> summary(aov(y~A+B+A:B,BlockData))
            Df   Sum Sq   Mean Sq   F value     Pr(>F)
A            2  108.170    54.085    9.4588   0.0029136 **
B            3  200.695    66.898   11.6997   0.0005388 ***
A:B          6    2.242     0.374    0.0653   0.9984654
Residuals   13   74.333     5.718
```

Observe that while main effects of the factor A and the blocking variable B are both significant, interaction effects are not. These results confirm observations from the earlier interaction plots.

Once it is determined that interaction effects are not significant, factor and blocking effects can be tested using

```
> summary(aov(y~A+B,BlockData))
            Df   Sum Sq   Mean Sq   F value     Pr(>F)
A            2  108.170    54.085   13.420    0.0002325 ***
B            3  200.695    66.898   16.599    1.537e-05 ***
Residuals   19   76.575     4.030
```

Since interaction is not significant, it should be expected that the results will not contradict the inferences obtained for the nonadditive model, which produces relevant significance tests for the additive model.

13.4   Diagnostics

Load the aov object linear model summary for the additive model using

```
Block.mod<-aov(y~A+B,BlockData)
Block.sum<-summary.lm(Block.mod)
```

Then gather the preliminary diagnostics statistics that might be needed.

```
e<-residuals(Block.mod)

y.hat<-fitted.values(Block.mod)

r<-e/(Block.sum$sigma*sqrt(1-hatvalues(Block.mod)))

d<-rstudent(Block.mod)
```

The methods used here almost mirror those used in diagnostics for earlier models.[3]

13.4.1   The constant variance assumption

Three plots can be looked at (see Figure 13.3). These plots are obtained using commands of the form:

```
with(BlockData,stripchart(r~A,method="stack",

        xlab="A",ylab=expression(hat(r)),vertical=T,las=1))

with(BlockData,stripchart(r~B,method="stack",

        xlab="B",ylab=expression(hat(r)),vertical=T,las=1))

plot(r~y.hat,xlab=expression(hat(y)),

                        ylab=expression(hat(r)),las=1)
```

Interpretations of the plots in Figure 13.3 remain as for earlier models.

The Brown–Forsyth test functions from earlier, including the function hov from package HH, can be used.

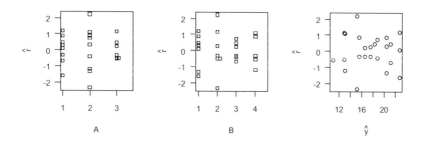

FIGURE 13.3: Plots of the studentized residuals against factor and block levels, and against the fitted values.

---

[3] The earlier caution on computing studentized and studentized deleted residuals for certain cases stands. Before using these residuals, it is a good idea to make sure that none are of the form NA or Inf.

For example, to use function hov, execute

```
> hov(e~A,BlockData)
        hov: Brown-Forsyth
data: e
F = 2.2081, df:A = 2, df:Residuals = 22, p-value = 0.1337
alternative hypothesis: variances are not identical
```

and

```
> hov(e~B,BlockData)
        hov: Brown-Forsyth
data: e
F = 0.5724, df:B = 3, df:Residuals = 21, p-value = 0.6394
alternative hypothesis: variances are not identical
```

The earlier constructed function Anova.bf.Ftest duplicates these results.

13.4.2   The normality assumption

As for earlier models,

```
qqnorm(r,main=NULL);abline(a=0,b=1,lty=3)
```

produces a QQ normal probability plot of the studentized residuals (see Figure 13.4), which serves the purpose.

The tests introduced in Chapter 6 and used for subsequent models still apply. For example,

```
> qq.cortest(r,0.05)
 QQ Normal Probability Corr. Coeff. Test, alpha = 0.05
data: r, n = 25
RQ = 0.9899, RCrit = 0.959
alternative hypothesis: If RCrit > RQ, Normality assumption
is invalid
```

and

```
> shapiro.test(r)
 Shapiro-Wilk normality test
data: r
W = 0.9839, p-value = 0.9498
```

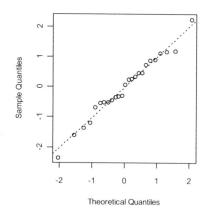

FIGURE 13.4: QQ normal probability plot of the studentized residuals with the line $y = x$ for reference.

13.4.3 The independence assumption

The same plots mentioned for use on earlier models, if applicable, serve to check the independence assumption. These include plots of $(\hat{\varepsilon}_i, \hat{\varepsilon}_{i+1})$, $i = 1, 2, \ldots, n - 1$, time-sequence plots, plots of the residuals against time, and plots of the residuals against variables not included in the model which might suggest any form of spatial or other correlation.

For the case of time-series data, the Durbin–Watson test can be used to test for the presence of first order autocorrelation.

13.4.4 The presence and influence of outliers

If all studentized deleted residuals for the model are well defined,[4] a plot of the studentized deleted residuals serves well. For example, in Figure 13.5 the use of Bonferroni cutoff values for the extreme 5% of cases identifies two outliers. Figure 13.5 was obtained using the commands

```
plot(d,ylab=expression(hat(d)))
m<-ceiling(.05*length(d));df<-Block.mod$df.residual-1
cv<-qt(0.05/(2*m),df,lower.tail=F)
abline(h=c(-cv,0,cv),lty=c(3,1,3))
identify(d)
```

---

[4] See caution in Section 11.4.

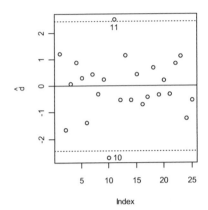

FIGURE 13.5: An index plot of the studentized deleted residuals along with Bonferroni cutoff lines computed for the most extreme 5% cases.

One might be interested in knowing whether Cases 10 and 11 are influential in any way. Influence measures also apply here, of particular interest being the DFBETAS because these relate to the parameters that may be tested or compared. As for the one-factor model in Chapter 12, the function influence.measures can be applied to Block.mod to look more closely at suspect cases.

13.5 Pairwise Comparisons of Treatment Effects

As for the one-factor model, consider the options of having a control treatment and of not having a control treatment.

13.5.1 With a control treatment

Since the function aov removes the first treatment (and blocking) parameter, suppose the first treatment (of factor A), is the control treatment. Then, it can be shown that for $i = 2, 3, \ldots, p$,

$$\mathrm{H}_0 : \beta_{A_i} = 0 \qquad \Longleftrightarrow \qquad \mathrm{H}_0 : \tau_{A_i} = \tau_{A_1}.$$

In this case,

```
> round(Block.sum$coefficients,4)
```

	Estimate	Std. Error	t value	Pr(>|t|)
(Intercept)	22.8503	0.9040	25.2781	0.0000
A2	-2.4444	0.9464	-2.5830	0.0182
A3	-4.6190	1.0395	-4.4435	0.0003

provides details on the necessary pairwise comparisons. That is,

Row A2 addresses $H_0 : \beta_{A_2} = 0$ or $\tau_{A_2} = \tau_{A_1}$; and

Row A3 addresses $H_0 : \beta_{A_3} = 0$ or $\tau_{A_3} = \tau_{A_1}$.

If a control is not present, or if all pairwise comparisons are desired for ranking purposes, earlier procedures find use.

13.5.2   Without a control treatment

The function TukeyHSD may be used.[5] Since the data were fitted to the model Block.mod using the formula y~A+B,

```
> TukeyHSD(Block.mod,"A",ordered=T)
 Tukey multiple comparisons of means
 95% family-wise confidence level
 factor levels have been ordered
Fit: aov(formula = y ~A + B, data = BlockData)
$A
```

	diff	lwr	upr	p adj
2-3	2.793651	0.22344566	5.363856	0.0318291
1-3	5.238095	2.66789010	7.808300	0.0001518
1-2	2.444444	0.04023768	4.848651	0.0458926

For the sake of it, compare these results with those obtained using

```
TukeyHSD(aov(y~B+A,BlockData),"A",ordered=T)
```

The code in Chapter 11 for the Tukey–Kramer procedure can be altered to duplicate these results.

---

[5] When using TukeyHSD on *unbalanced designs* involving more than one factor, always enter the model formula with the factor of interest coming first; that is, if A represents the factor of interest, then enter the formula in the form y ~A + B. This outputs results that correctly report test results of pairwise differences for factor A.

13.6 Tukey's Nonadditivity Test

It is possible that constraints on a study may limit data collection to exactly one observed response per cell. The only difference between this scenario and the earlier one is that the F-test for interaction cannot be performed on the nonadditive model. One method of determining if interaction effects might be present is to use interaction plots, as shown earlier. Another is to use what is referred to as *Tukey's Nonadditivity Test*.[6] Only assessments for interaction effects are discussed here as once interaction is determined not to be significant the process remains as for the previous case.

Denote the continuous response variable by Y and the two categorical variables by A and B, where B is the blocking variable. Consider the data shown in Table 13.2.

TABLE 13.2: Data with One Observation per Cell, to Illustrate Tukey's Nonadditivity Test.

	B1	B2	B3	B4
A1	35	15	28	19
A2	22	5	15	10
A3	29	5	31	15

The data are stored in `Data13x02.R` as a data frame named `Block2Data`. As before, the variable A is assigned values 1, 2, and 3, and values for the variable B are 1, 2, 3, and 4. Remember to make sure such variables are defined as factors.

Interaction plots for this dataset are shown in Figure 13.6. One might suspect interaction since the two plots do not appear to show (approximate) vertical translations of the traces.

Tukey's Nonadditivity Test looks at testing

$$H_0 : y_{ijk} = \mu + \tau_{A_i} + \tau_{B_j} + \varepsilon_{ijk} \quad \text{against}$$
$$H_1 : y_{ijk} = \mu + \tau_{A_i} + \tau_{B_j} + \delta \tau_{A_i} \tau_{B_j} + \varepsilon_{ijk}$$

where the parameter δ is an unknown constant representing interaction effects.

The strategy (using R) is first to fit the additive model, and then, using parameter estimates from the null (additive) model, construct the alternative (nonadditive) model and compare the two. Recall that the default treatment

---

[6] Also called Tukey's Additivity Test!

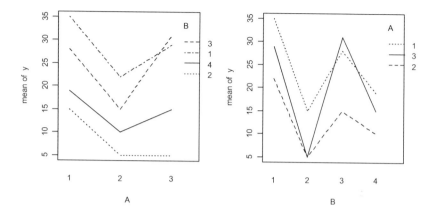

FIGURE 13.6: Interaction plots for dataset `Block2Data` to determine whether interaction between the levels of the factor and blocking variable exist.

coding scheme has been used thus far. Continuing with this scheme, the models being compared are

$$H_0 : y_{ijk} = \beta_0 + \beta_{A_i} + \beta_{B_j} + \varepsilon_{ijk} \quad \text{against}$$
$$H_1 : y_{ijk} = \beta_0 + \beta_{A_i} + \beta_{B_j} + \delta \beta_{A_i} \beta_{B_j} + \varepsilon_{ijk},$$

with $\beta_{A_1} = \beta_{B_1} = 0$ being predetermined by the treatment coding scheme.

```
#Fit the additive model
add.mod<-lm(y~A+B,Block2Data)
#Extract factor coefficient estimates
A.C<-rep(c(0,add.mod$coefficients[2:3]),each=4)
#Extract block coefficient estimates
B.C<-rep(c(0,add.mod$coefficients[4:6]),3)
#Fit the nonadditive model
nonadd.mod<-lm(y~A+B+I(A.C*B.C),Block2Data)
#Compare models
anova(add.mod,nonadd.mod)
```

and get

```
Analysis of Variance Table
Model 1: y ~A + B
```

```
Model 2: y ~A + B + I(A.C * B.C)
  Res.Df     RSS  Df  Sum of Sq      F  Pr(>F)
1       6  78.500
2       5  69.783   1     8.7166  0.6245  0.4652
```

So, there is insufficient evidence to conclude nonadditivity.

Some explanation might be appropriate for the construction of the objects A.C and B.C. In forming A.C, since there are four block levels and the data are stored by levels of factor A (first all of level 1, then level 2, and finally level 3), the parameter estimates for *each* of the factor levels $(\hat{\beta}_{A_1} = 0, \hat{\beta}_{A_2}, \hat{\beta}_{A_3})$ are repeated four times (one for each blocking level). Next, for B.C, since there are three treatments the parameter estimates for the block levels $(\hat{\beta}_{B_1} = 0, \hat{\beta}_{B_2}, \hat{\beta}_{B_3}, \hat{\beta}_{B_4})$ are iterated three times (once for each factor level).

13.7 For the Curious

The dataset BlockData is used directly in the following illustration. Also given in this section is an illustration of how simulated data may be generated for block designs and used in practicing interpreting interaction plots.

13.7.1 Bonferroni's pairwise comparisons

If g represents the number of simultaneous pairwise tests being conducted, Bonferroni's procedure can be performed by preparing code to compute interval bounds and p-values using

$$\left(\hat{\mu}_{A_l} - \hat{\mu}_{A_m}\right) \pm t\left(\alpha/(2g), n - I - J + 1\right) s \sqrt{\frac{1}{n_{A_l}} + \frac{1}{n_{A_m}}},$$

and

$$T_{lm} = \frac{\left|\hat{\mu}_{A_l} - \hat{\mu}_{A_m}\right|}{s\sqrt{1/n_{A_l} + 1/n_{A_m}}}.$$

Sample code to perform these calculations is shown as follows.

```
#For convenience
attach(BlockData)
#Initialize counters
n<-length(y);I<-length(levels(A));J<-length(levels(B))
g<-choose(I,2)
```

```
#Get cell sizes and means
nA<-tapply(y,A,length);means<-tapply(y,A,mean)
detach(BlockData)
#Initialize storage variables
Pair<-character(g);Diff<-numeric(g)
S.er<-numeric(g)
Tv<-numeric(g);lwr<-numeric(g)
upr<-numeric(g);Pv<-numeric(g)
#Get critical value for intervals
cv<-qt(.05/(2*g),n-I-J+1,lower.tail=F)
#Initialize test counter and begin for-loops
k<-1
for (l in 1:(I-1)){
    for (m in (l+1):I){
        #Perform calculations
        Pair[k]<-paste("A",l," vs. ","A",m,sep="")
        Diff[k]<-means[l]-means[m]
        S.er[k]<-Block.sum$sigma*sqrt(1/nA[l]+1/nA[m])
        lwr[k]<-round(Diff[k]-cv*S.er[k],4)
        upr[k]<-round(Diff[k]+cv*S.er[k],4)
        Tv[k]<-abs(Diff[k])/S.er[k]
        Pv[k]<-round(2*g*pt(Tv[k],n-I-J+1,lower.tail=F),3)
        Pv[k]<-ifelse(Pv[k]>1,1,Pv[k]);k<-k+1}}
#Store results
Results<-data.frame(Diff,S.er,lwr,upr,Pv,row.names=Pair)
```

The results are

```
> Results
               Diff      S.er      lwr      upr      Pv
A1 vs. A2  2.444444  0.9463704  -0.0399   4.9288   0.055
A1 vs. A3  5.238095  1.0117125   2.5822   7.8939   0.000
A2 vs. A3  2.793651  1.0117125   0.1378   5.4495   0.037
```

Observe that Scheffes's procedure can be implemented simply by replacing

```
cv<-qt(.05/(2*g),n-I-J+1,lower.tail=F)   by
cv<-sqrt((I-1)*qf(.05,(I-1),n-I-J+1,lower.tail=F))
```

then

```
Tv[k]<-abs(Diff[k])/S.er[k]   by
Tv[k]<-(Diff[k])^2/((I-1)*S.er[k]^2)
```

and, finally,

```
Pv[k]<-round(2*g*pt(Tv[k],n-I-J+1,lower.tail=F),3)   by
Pv[k]<-round(pf(Tv[k],I-1,n-I-J+1,lower.tail=F),3)
```

in the above code for Bonferroni's procedure.

13.7.2   Generating data to play with

For $i = 1, 2, \ldots, I$ and $j = 1, 2, \ldots, J$, respectively, the complete interaction effects model has the appearance

$$y_{ijk} = \mu + \tau_{A_i} + \tau_{B_j} + \tau_{AB_{ij}} + \varepsilon_{ijk}.$$

It is helpful to have a feel for what interaction plots might look like under different conditions. Getting data that illustrate the various scenarios is not always convenient, but generating such data is quite easy with R.

```
#Decide on the number of treatments and blocks
I<-3;J<-4
#Set bound for effects parameters
a<-.1;b<-.5;ab<-.1
#Get an overall mean for the response
mu<-runif(1,25,30)
#Generate treatment and blocking effects values
TaoA<-runif(I,-a,a);TaoB<-runif(J,-b,b)
#Generate interaction effects values
TaoAB<-matrix(runif(I*J,-ab,ab),nrow=I,ncol=J)
#Set cell sizes
nij<-matrix(rbinom(I*J,12,.5)+1,nrow=I,ncol=J)
#Get total number of observations
n<-sum(nij)
```

```
#Initialize data variables
y<-numeric(n);A<-integer(n);B<-integer(n)
#Get a random sample of error terms
e<-rnorm(n,mean=0,sd=.1)
#Initialize variable index
l<-1
#Start data creation loops
for (i in 1:I){
    for (j in 1:J){
        for (k in 1:nij[i,j]){
        #Get values for the ij cell
        A[l]<-i;B[l]<-j
        y[l]<-mu+TaoA[i]+TaoB[j]+TaoAB[i,j]+e[k]
        #Update variable index and close loops
        l<-l+1}}}
#Store data in a data frame
TestData<-data.frame(y,A,B)\
#Define treatment and blocking variables as factors
TestData$A<-factor(TestData$A)
TestData$B<-factor(TestData$B)
```

Then commands of the form

```
#Get interaction plots
win.graph(width=6,height=3.5)
par(mfrow=c(1,2),cex=.75,ps=10)
with(TestData,interaction.plot(A,B,y))
with(TestData,interaction.plot(B,A,y))
par(mfrow=c(1,1))
#Get ANOVA table for nonadditive model
summary(aov(y~A+B+A:B,TestData))
```

can be used to look at the appearance of interaction plots along with the corresponding goodness-of-fit F-test (ANOVA table).

To create a dataset with exactly one observation per cell, simply use

```
nij<-matrix(rep(1,I*J),nrow=I,ncol=J)
```

to set the cell size in the above code.

The appearance of the interaction plots and the results in the corresponding ANOVA table will change for each run of the code, and more drastic changes can be encouraged by adjusting the values assigned to a, b and ab.

Chapter 14

Two-Factor Models

14.1 Introduction 319
14.2 Exploratory Data Analysis 321
14.3 Model Construction and Fit 322
14.4 Diagnostics 324
14.5 Pairwise Comparisons of Treatment Effects 325
    14.5.1 With a control treatment 326
    14.5.2 Without a control treatment 327
14.6 What if Interaction Effects Are Significant? 328
14.7 Data with Exactly One Observation per Cell 331
14.8 Two-Factor Models with Covariates 331
14.9 For the Curious: Scheffe's F-Tests 331

14.1 Introduction

Presentations in this chapter are briefer than earlier chapters since much of what was done for the block designs in Chapter 13, along with earlier material, remains applicable.[1]

Consider an experiment involving two *fixed-effects factors*; call these factors A and B. Suppose factor A has I levels and B has J levels. Let y_{ijk} represent the kth randomly observed response for the ith level of factor A and j^{th} level of B, or the kth observation in the ij^{th} cell. Let n_{ij} represent the number of observations within the ij^{th} cell and $n = \sum_{i=1}^{I} \sum_{j=1}^{J} n_{ij}$ the total number of observed responses.

Denote each cell mean by the unknown parameter μ_{ij}; then, for $i = 1, 2, \ldots, I$, $j = 1, 2, \ldots, J$, and $k = 1, 2, \ldots, n_{ij}$ each observed response y_{ijk} may be expressed as the sum of the corresponding cell mean and a random error term ε_{ijk}, giving the *cell means model*,

$$y_{ijk} = \mu_{ij} + \varepsilon_{ijk}.$$

Alternatively, let μ denote the overall mean response; then decompose the cell means into the sum of *main effects parameters*, τ_{A_i} and τ_{B_j}, and *interaction effects parameters*, $\tau_{AB_{ij}}$,

$$\mu_{ij} = \mu + \tau_{A_i} + \tau_{B_j} + \tau_{AB_{ij}}.$$

---

[1] Be sure to reset the contrast to the defaults setting to duplicate results in this chapter. An easy way to do this is simply to restart R.

Substitution of this expression in the cell means model yields the *nonadditive effects model,*

$$y_{ijk} = \mu + \tau_{A_i} + \tau_{B_j} + \tau_{AB_{ij}} + \varepsilon_{ijk}.$$

If the interaction effects are absent, the appropriate model is the *additive effects model*

$$y_{ijk} = \mu + \tau_{A_i} + \tau_{B_j} + \varepsilon_{ijk}.$$

All three models mentioned above have a matrix representation,

$$\mathbf{y} = \mathbf{X}\boldsymbol{\beta} + \boldsymbol{\varepsilon},$$

with terms being as previously defined. Only the cell means model has a design matrix with full column rank.

Assumptions for the models discussed in this chapter are as for earlier models. The error terms ε_{ijk} are independent, and are identically and normally distributed with $\varepsilon_{ijk} \sim N(0, \sigma^2)$.

Data for designs involving two fixed-effects factors are coded in the same manner as for block designs discussed in the previous chapter. This chapter addresses only complete designs, those for which every cell contains at least one observed response. In practice, the preference would be complete designs in which at least one cell has more than one observation.

Two illustrations are given: One for which the data are appropriate for an additive model (Table 14.1) and the other for which the data are appropriate for a nonadditive model (Table 14.2). Both datasets were generated using code of the form given in Section 13.7.2.

TABLE 14.1: Data for An Additive Model Illustration

	B1	B2	B3	B4
A1	7.43	10.92	9.07	8.71
	7.50	11.04	8.88	8.55
	7.50	11.05	8.96	8.72
			9.06	8.42
A2	5.77	9.56	7.36	6.81
	5.67	9.49	7.50	6.95
	5.82	9.45	7.67	7.05
	5.92	9.53	7.49	6.87
		9.33	7.40	7.13
A3	6.28	9.68	7.92	7.25
	6.28	9.74	7.87	7.39
	6.31	9.80		7.68
	6.28			
	6.28			
	6.30			

TABLE 14.2: Data for An Illustration of a Nonadditive Model

	B1	B2	B3	B4
A1	10.78	10.21	11.66	11.70
	10.48	10.41	11.60	11.42
	10.65		11.83	11.62
	10.77			
A2	10.35	10.01	11.72	11.05
	10.58	9.58	12.13	11.10
	10.23		11.69	
	10.51		11.97	
	10.59			
A3	10.17	10.42	11.48	10.66
	10.55	10.42	11.00	11.52
	10.55		11.36	10.98
	10.36			11.11
	10.40			11.10
				10.99

Begin with the additive model. The data in Table 14.1 are stored in `Data14x01.R` as a data frame named `TwoWayAdd` with the variable A being assigned values 1, 2, and 3, and for the variable B, values assigned are 1, 2, 3, and 4. Once the data are sourced, make sure categorical variables are defined as factors.

14.2 Exploratory Data Analysis

As with the earlier block design, interaction plots for the dataset `TwoWayAdd` serve as good first checks for the presence of interaction between the levels of the two factors. From Figure 14.1 it is evident that interaction is most likely not significant in the data since the traces in each of the interaction plots appear to be approximate vertical translations of each other. Moreover, both figures suggest that main effects for both factors are most likely significant.

Data summaries can be obtained as earlier using the functions `summary`, `by`, and `tapply`. Similarly, a visual backup of the numeric summaries can be performed with the help of figures such as those in Figure 14.2. Code for Figures 14.1 and 14.2 are as for Figures 13.1 and 13.2 for the block design data in Chapter 13.

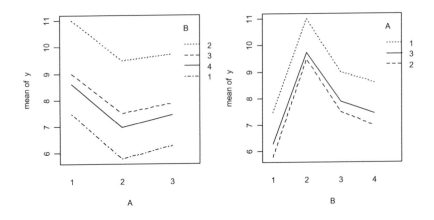

FIGURE 14.1: Interaction plots for dataset `TwoWayAdd` to determine the possible presence of interaction between levels of factor A and factor B.

14.3 Model Construction and Fit

Unless there is very strong evidence to support the use of an additive model, theoretical or otherwise, the nonadditive model is typically assumed first, which in unadjusted form has the appearance

$$y_{ijk} = \mu + \tau_{A_i} + \tau_{B_j} + \tau_{AB_{ij}} + \varepsilon_{ijk},$$

along with appropriate conditions on the effects and interaction parameters. See, for example, [45, p. 832 and p. 953]. Ideas used for adjusting the one-factor effects model and the block model extend to the two-factor case. By default, R uses treatment coding to generate the equivalant (adjusted) full rank model. This is accomplished by assuming an equivalent model of the form

$$y_{ijk} = \beta_0 + \beta_{A_i} + \beta_{B_j} + \beta_{AB_{ij}} + \varepsilon_{ijk},$$

where $i = 1, 2, \ldots, I$, $j = 1, 2, \ldots, J$, and $k = 1, 2, \ldots, n_{ij}$, and subject to $\beta_{A_1} = 0$, $\beta_{B_1} = 0$, and $\beta_{AB_{1j}} = \beta_{AB_{i1}} = 0$ for $i = 1, 2, \ldots, I$, $j = 1, 2, \ldots, J$. Similarly, the adjusted additive model has the appearance

$$y_{ijk} = \beta_0 + \beta_{A_i} + \beta_{B_j} + \varepsilon_{ijk},$$

where $i = 1, 2, \ldots, I$, $j = 1, 2, \ldots, J$, and $k = 1, 2, \ldots, n_{ij}$, with $\beta_{A_1} = 0$, $\beta_{B_1} = 0$.

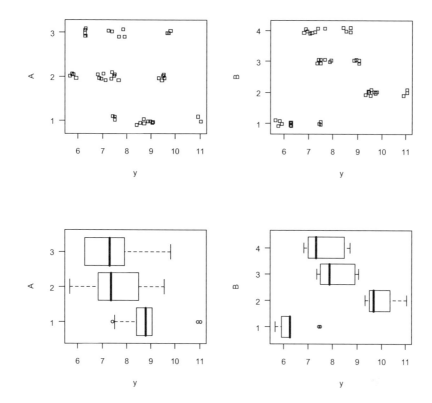

FIGURE 14.2: Stripcharts and boxplots of the response variable data in the dataset `TwoWayAdd` by factor levels.

A quick check for interaction can be accomplished using

```
> summary(aov(y~A+B+A:B,TwoWayAdd))
          Df  Sum Sq  Mean Sq    F value  Pr(>F)
A          2  21.428  10.7141   950.5004  <2e-16 ***
B          3  76.255  25.4184  2254.9907  <2e-16 ***
A:B        6   0.085   0.0142     1.2585  0.3014
Residuals 35   0.395   0.0113
```

which suggests the interaction effects are not significant.

Assuming the additive model is the way to go, the necessary model objects are first obtained.

```
TwoWayAdd.mod<-aov(y~A+B,TwoWayAdd)
TwoWayAdd.sum<-summary.lm(TwoWayAdd.mod)
```

and

```
> summary(TwoWayAdd.mod)
             Df  Sum Sq  Mean Sq   F value       Pr(>F)
A             2  21.428  10.7141    915.85   < 2.2e-16 ***
B             3  76.255  25.4184   2172.79   < 2.2e-16 ***
Residuals    41   0.480   0.0117
```

indicates, as with the nonadditive model, that both main effects are significant.

14.4 Diagnostics

As before, first gather the necessary statistics[2] using commands of the form:

```
e<-residuals(TwoWayAdd.mod)
y.hat<-fitted.values(TwoWayAdd.mod)
r<-e/(TwoWayAdd.sum$sigma*sqrt(1-hatvalues(TwoWayAdd.mod)))
d<-rstudent(TwoWayAdd.mod)
```

Then go through the list of checks as with models in previous chapters. Figure 14.3 provides all diagnostic plots for the error assumptions and Figure 14.4 permits a quick check for the presence of outliers among the most extreme 5% of studentized deleted residuals. These figures were obtained using appropriately altered code from Section 13.4. Assessing for possible violations of the independence assumption remains as for earlier models.

The Brown–Forsyth test, the QQ normal probability correlation coefficient test, and the Shapiro–Wilk test all apply with appropriately altered commands from Section 13.5. It will be found that at $\alpha = 0.05$, there is insufficient evidence to reject the constant variances[3] and normality assumptions on the error terms.

Since Figure 14.4 indicates the presence of an outlier, Case 47, it might be appropriate to look at the influence measures for this case. Using the function `influence.measures`, it will be found that Case 47 is not influential on any of the parameter estimates. Moreover, the Cook's distance suggests this case is not influential on all predicted mean responses and, consequently, it may be assumed that there is probably not much to worry about here.

---

[2] The earlier caution for studentized residuals and studentized deleted residuals applies. If any of the leverages equal one, do not use these two forms of residuals.

[3] Use either of the functions `hov` from package `HH` or `Anova.bf.Ftest` from Chapter 11. The assumption of constant variances across levels of factor B appears quite marginal.

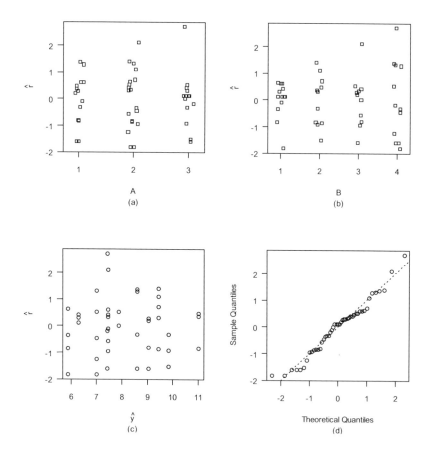

FIGURE 14.3: Diagnostic plots to assess the constant variance and normality assumptions of the error terms for the additive model `TwoWayAdd.mod`.

If applicable, the Durbin–Watson test can be used to assess for the possible presence of autocorrelation.

14.5 Pairwise Comparisons of Treatment Effects

As for the earlier block design, it can be shown that if the first treatment in a factor is the control treatment, then the following equivalences can be shown: For $i = 2, \ldots, I$,

$$\text{H}_0 : \beta_{A_i} = 0 \iff \text{H}_0 : \tau_{A_i} = \tau_{A_1};$$

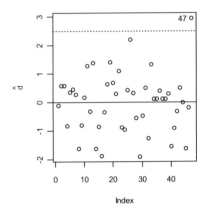

FIGURE 14.4: Plot of studentized deleted residuals with Bonferroni cutoff lines for the most extreme 5% of plotted points.

and for $j = 2, \ldots, J$,

$$\text{H}_0 : \beta_{B_j} = 0 \quad \Longleftrightarrow \quad \text{H}_0 : \tau_{B_j} = \tau_{B_1}.$$

Similarly, for the remaining pairwise comparisons, it can be shown that for $i, j = 2, \ldots, I$ with $i \neq j$,

$$\text{H}_0 : \beta_{A_i} = \beta_{A_j} \quad \Longleftrightarrow \quad \text{H}_0 : \tau_{A_i} = \tau_{A_j};$$

and for $l, m = 2, \ldots, J$ with $l \neq m$,

$$\text{H}_0 : \beta_{B_l} = \beta_{B_m} \quad \Longleftrightarrow \quad \text{H}_0 : \tau_{B_l} = \tau_{B_m}.$$

Thus, the `summary.lm` object provides one way of performing pairwise comparisons when the first treatment of each factor is the control.

14.5.1 With a control treatment

If the first treatment in each factor serves as a control, the results of pairwise comparisons are given in

```
> TwoWayAdd.sum$coefficients
              Estimate  Std. Error    t value       Pr(>|t|)
(Intercept)   7.439154  0.04058959  183.27737   2.274567e-61
A2           -1.584498  0.03814345  -41.54051   3.921245e-35
A3           -1.171834  0.04174978  -28.06802   2.173160e-28
```

B2	3.563391	0.04483658	79.47508	1.546087e-46
B3	1.601406	0.04541072	35.26494	2.702871e-32
B4	1.141513	0.04394667	25.97495	4.417804e-27

The output suggests all treatment effects parameters in factor A differ from τ_{A_1}, and all treatment effects parameters in factor B differ from τ_{B_1}.

Once again, the routine glht from package multcomp can also be used to duplicate the above tests, perform other pairwise comparisons, or test any other general contrast within each factor.

14.5.2 Without a control treatment

Tukey's procedure can be used here, too; however, recall the caution from the previous chapter.[4] To perform pairwise tests for levels of factor A, enter

```
TukeyHSD(aov(y~A+B,TwoWayAdd),"A",ordered=T)
```

to get

```
$A
```

	diff	lwr	upr	p adj
2-3	0.009924812	-0.08271182	0.1025614	0.9633107
1-3	1.482142857	1.38273572	1.5815500	0.0000000
1-2	1.472218045	1.37958142	1.5648547	0.0000000

and to perform pairwise comparisons for factor B, enter

```
TukeyHSD(aov(y~B+A,TwoWayAdd),"B",ordered=T)
```

to get

```
$B
```

	diff	lwr	upr	p adj
4-1	1.2167308	1.1007938	1.332668	0
3-1	1.6965035	1.5778578	1.815149	0
2-1	3.5519580	3.4333124	3.670604	0
3-4	0.4797727	0.3588825	0.600663	0
2-4	2.3352273	2.2143370	2.456118	0
2-3	1.8554545	1.7319642	1.978945	0

---

[4]This caution applies to unbalanced designs – see footnote on p. 311.

Code for Bonferroni's procedure can be prepared by tacking computations for g comparisons within factor B onto the code prepared in Section 13.7. The additional computations needed are

$$\left(\hat{\mu}_{B_l} - \hat{\mu}_{B_m}\right) \pm t\left(\alpha/(2g), n - I - J + 1\right) s \sqrt{\frac{1}{n_{B_l}} + \frac{1}{n_{B_m}}}$$

for the confidence intervals, and

$$T_{lm}^* = \frac{\left|\hat{\mu}_{B_l} - \hat{\mu}_{B_m}\right|}{s \sqrt{1/n_{B_l} + 1/n_{B_m}}}$$

for the test values. This can be accomplished simply by altering the earlier code to handle the additional factor. The results show

```
              Diff        S.er     lwr     upr  Pv
A1 vs. A2  1.472218045  0.03809619  1.3771  1.5673   0
A1 vs. A3  1.482142857  0.04088051  1.3801  1.5842   0
A2 vs. A3  0.009924812  0.03809619 -0.0852  0.1050   1
```

and

```
              Diff        S.er      lwr      upr   Pv
B1 vs. B2  -3.5519580  0.04431012  -3.6748  -3.4291   0
B1 vs. B3  -1.6965035  0.04431012  -1.8193  -1.5737   0
B1 vs. B4  -1.2167308  0.04329852  -1.3368  -1.0967   0
B2 vs. B3   1.8554545  0.04611944   1.7276   1.9833   0
B2 vs. B4   2.3352273  0.04514839   2.2101   2.4604   0
B3 vs. B4   0.4797727  0.04514839   0.3546   0.6049   0
```

Sample code to implement Scheffe's pairwise comparisons is given in the last section of this chapter.

14.6 What if Interaction Effects Are Significant?

The data in Table 14.2 are stored in `Data14x02.R` as a data frame named `TwoWayInt`. Again, the variable A is assigned values 1, 2, and 3, and values for the variable B are 1, 2, 3, and 4.

The interaction plots (see Figure 14.5) for these data suggest the presence of interaction.

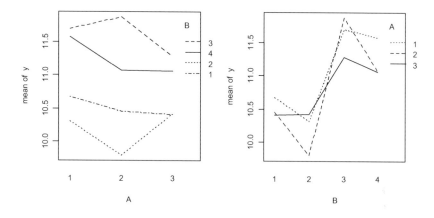

FIGURE 14.5: Interaction plots for the dataset `TwoWayInt`, a dataset in which the data exhibit interactions between factor levels.

This observation is supported by

```
> summary(aov(y~A+B+A:B,TwoWayInt))
```

	Df	Sum Sq	Mean Sq	F value	Pr(>F)	
A	2	0.5497	0.2749	7.3519	0.002614	**
B	3	12.0735	4.0245	107.6429	8.089e-16	***
A:B	6	1.1564	0.1927	5.1549	0.001032	**
Residuals	29	1.0842	0.0374			

Note that some residual analysis for the fitted model should probably be performed. However, suppose no assumptions are violated.

It is suggested by some that if interactions between factor levels are significant, then the data can be fitted to a one-factor model in which all the cells are viewed as treatments within a single (larger) factor.

Consider the dataset `TwoWayInt` to illustrate this. It would be helpful to create a new categorical variable, call this `AB`, and assign values of $11, 12, 13, 14, \ldots, ij, \ldots$, where $i = 1, 2, 3$ and $j = 1, 2, 3, 4$.

```
#For convenience
attach(TwoWayInt)
#Initialize the cell variable
AB<-character(length(y))
#Store values in AB
for (i in 1:length(y)){AB[i]<-paste(A[i],B[i],sep="")}
```

```
#Create a new data frame
IntMod.dat<-data.frame(y,AB); rm(AB)
#Define AB as a factor, to be safe
IntMod.dat$AB<-factor(IntMod.dat$AB)
detach(TwoWayInt)
```

Then

```
> summary(aov(y~AB,IntMod.dat))
            Df   Sum Sq  Mean Sq  F value     Pr(>F)
AB          11  13.7797  1.25270   33.506  1.647e-13 ***
Residuals   29   1.0842  0.03739
```

Tukey's pairwise comparisons may be performed to isolate differences; however, there is an inconvenience in this situation. Since there are 12 cells, the number of pairwise comparisons is choose(12,2) = 66. It would be a nuisance to go through so many tests one by one. While a plot of Tukey's tests can be obtained, for example, try (plot not shown):

```
tests<-TukeyHSD(aov(y~AB,IntMod.dat))
plot(tests)
```

this plot is not easy to navigate because it contains too much information. If one were interested only in pairs that are not significantly different, another approach would be to make use of a for-loop. For example, the code

```
attach(tests); j<-0
for (i in 1:66){
    if (AB[i,4]>=0.05) {j<-j+1}}
detach(tests);j
```

counts the number of pairs for which the effects parameters are not significantly different at $\alpha = 0.05$; it turns out there are 25 such pairs. To identify those pairs that are not significantly different at $\alpha = 0.05$, one might execute code of the form:

```
attach(tests)
H.Pairs<-NULL;
for (i in 1:66){
  if (AB[i,4]>=0.05){
      H.Pairs<-c(H.Pairs,row.names(AB)[i])}}
detach(tests);H.Pairs
```

which lists all such pairs. An examination of the contents of H.Pairs would then reveal homogeneous subsets. While a program can be written to perform this task, careful planning and coding would be required.

14.7   Data with Exactly One Observation per Cell

As for the block design illustration in Chapter 13, it is possible that constraints on a study may limit data collection to exactly one observed response per cell. Refer to the previous chapter for an illustration of *Tukey's Nonadditivity Test*. The process is identical for two-factor models.

14.8   Two-Factor Models with Covariates

The process is analogous to the one-factor case. The additive model in the default setting has the appearance

$$y_{ijk} = \beta_0 + \beta_{A_i} + \beta_{B_j} + \beta_1 x_{ijk} + \varepsilon_{ijk}$$

and the data are fitted to the model using

```
lm(y~A+B+x,data.set)
```

If there are two or more covariates, the command takes on the appearance

```
lm(y~A+B+x1+x2+···+xq,data.set)
```

The constant slope assumption is tested in an analogous manner, with the added complication of having to take both factors into consideration.

14.9   For the Curious: Scheffe's F-Tests

With respect to performing Scheffe's procedure on the additive two-factor model, much of what is discussed in the last section of the previous chapter applies, the difference being that for the current illustration both factors are of interest.

For testing pairs of factor A main effects, bounds for confidence intervals are computed using

$$\left(\hat{\mu}_{A_l} - \hat{\mu}_{A_m}\right) \pm s \sqrt{(I-1)\, F\left(\alpha, I-1, n-I-J+1\right) \left(\frac{1}{n_{A_l}} + \frac{1}{n_{A_m}}\right)},$$

and test values are obtained using

$$F^*_{lm} = \frac{\left(\hat{\mu}_{A_l} - \hat{\mu}_{A_m}\right)^2}{s^2 \left(I - 1\right) \left(\dfrac{1}{n_{A_l}} + \dfrac{1}{n_{A_m}}\right)}.$$

Analogously, for testing pairs of factor B main effects, bounds for confidence intervals are obtained using

$$\left(\hat{\mu}_{B_l} - \hat{\mu}_{B_m}\right) \pm s \sqrt{\left(J - 1\right) F\left(\alpha, J - 1, n - I - J + 1\right) \left(\frac{1}{n_{B_l}} + \frac{1}{n_{B_m}}\right)},$$

and the test values by

$$F^*_{lm} = \frac{\left(\hat{\mu}_{B_l} - \hat{\mu}_{B_m}\right)^2}{s^2 \left(J - 1\right) \left(\dfrac{1}{n_{B_l}} + \dfrac{1}{n_{B_m}}\right)};$$

Here is sample code for Scheffe's procedure as applied to the additive two-factor model.

```
#For convenience
attach(TwoWayAdd)
#Gather various lengths
n<-length(y); I<-length(levels(A))
J<-length(levels(B))
#Obtain number of pairwise comparisons
gA<-choose(I,2);gB<-choose(J,2)
#Obtain cell sizes
nA<-tapply(y,A,length);nB<-tapply(y,B,length)
#Obtain cell means
meansA<-tapply(y,A,mean);meansB<-tapply(y,B,mean)
detach(TwoWayAdd)
#Initialize various variables that will be needed
PairA<-character(gA);PairB<-character(gB)
DiffA<-numeric(gA);DiffB<-numeric(gB)
S.erA<-numeric(gA);S.erB<-numeric(gB)
TvA<-numeric(gA);TvB<-numeric(gB)
lwrA<-numeric(gA);lwrB<-numeric(gB)
uprA<-numeric(gA);uprB<-numeric(gB)
```

```
PvA<-numeric(gA);PvB<-numeric(gB)
#Obtain "critical values"
cvA<-sqrt((I-1)*qf(.05,(I-1),n-I-J+1,lower.tail=F))
cvB<-sqrt((J-1)*qf(.05,(J-1),n-I-J+1,lower.tail=F))
#Initialize comparison counter
k<-1
#Start computing details for factor A
for (l in 1:2){for (m in (l+1):3){
    #Identify pair, obtain signed difference in means
    PairA[k]<-paste("A",l," vs. ","A",m,sep="")
    DiffA[k]<-meansA[l]-meansA[m]
    #Obtain standard error
    S.erA[k]<-TwoWayAdd.sum$sigma*sqrt(1/nA[l]+1/nA[m])
    #Compute lower and upper bounds and test values
    lwrA[k]<-DiffA[k]-cvA*S.erA[k]
    uprA[k]<-DiffA[k]+cvA*S.erA[k]
    TvA[k]<-(DiffA[k])^2/((I-1)*S.erA[k]^2)
    #Get p-values and close the loop
    PvA[k]<-round(pf(TvA[k],I-1,n-I-J+1,lower.tail=F),3)
    PvA[k]<-ifelse(PvA[k]>1,1,PvA[k])
    k<-k+1}}
#Store the results
ScheffeA<-data.frame(cbind(DiffA,S.erA,lwrA,uprA,PvA),
                                        row.names=PairA)
#Repeat the process for factor B
k<-1
for (l in 1:3){for (m in (l+1):4){
PairB[k]<-paste("B",l," vs. ","B",m,sep="")
DiffB[k]<-meansB[l]-meansB[m]
S.erB[k]<-TwoWayAdd.sum$sigma*sqrt(1/nB[l]+1/nB[m])
lwrB[k]<-DiffB[k]-cvA*S.erB[k]
uprB[k]<-DiffB[k]+cvB*S.erB[k]
TvB[k]<-(DiffB[k])^2/((J-1)*S.erB[k]^2)
PvB[k]<-round(pf(TvB[k],J-1,n-I-J+1,lower.tail=F),3)
```

```
PvB[k]<-ifelse(PvB[k]>1,1,PvB[k])
k<-k+1}}
ScheffeB<-data.frame(cbind(DiffB,S.erB,lwrB,uprB,PvB),
                                        row.names=PairB)
```

Then

```
> list("A"=round(ScheffeA,3),"B"=round(ScheffeB,3))
$A
```

	DiffA	S.erA	lwrA	uprA	PvA
A1 vs. A2	1.472	0.0381	1.375	1.569	0.000
A1 vs. A3	1.482	0.0409	1.378	1.586	0.000
A2 vs. A3	0.010	0.0381	-0.087	0.107	0.967

```
$B
```

	DiffB	S.erB	lwrB	uprB	PvB
B1 vs. B2	-3.552	0.0443	-3.665	-3.423	0
B1 vs. B3	-1.697	0.0443	-1.809	-1.567	0
B1 vs. B4	-1.217	0.0433	-1.327	-1.091	0
B2 vs. B3	1.855	0.0461	1.738	1.990	0
B2 vs. B4	2.335	0.0451	2.221	2.467	0
B3 vs. B4	0.480	0.0451	0.365	0.611	0

This program can be refined to produce better looking output and can also be turned into a function that accepts any additive two-factor model. Note also that this code can be adapted to work for block designs as well.

Chapter 15

Simple Remedies for Fixed-Effects Models

15.1 Introduction 335
15.2 Issues with the Error Assumptions 335
15.3 Missing Variables 336
15.4 Issues Specific to Covariates 336
    15.4.1 Multicollinearity 336
    15.4.2 Transformations of covariates 336
    15.4.3 Blocking as an alternative to covariates 337
15.5 For the Curious 339

15.1 Introduction

Most of the remedial measures mentioned in Chapters 7 and 10 apply to models discussed in Chapters 11 through 14. Consequently, the corresponding computational methods described in Chapters 7 and 10 also apply to models covered in Chapters 11 through 14. This chapter first recalls those methods used for regression models that are applicable to one- and two-factor models with fixed-effects factors; this includes models involving blocking variables and covariates. Methods specific to models that include covariates wrap up this Companion.

15.2 Issues with the Error Assumptions

Transformations of the response variable as suggested by the literature, determined through observation or obtained by numerical methods, all apply to the models under discussion and are implemented as for regression models. When simple remedies do not work to resolve nonconstant variance issues, weighted least squares can be employed. If transformations do not help resolve issues with the normality or constant variance assumptions, one recourse is to use nonparametric methods; for example, see [22] and [45].

For matters relating to the independence of the error terms, it is suggested that care be exercised in the design phase of experiments to ensure this does not become an issue in the first place. Time series data are typically handled by methods that accommodate autocorrelation. These methods are beyond the scope of this Companion; see, for example, [10], [14], and [16] for more on the subject.

15.3   Missing Variables

As with regression models, the absence of important variables, numeric or categorical, can give rise to issues. The computational fix is simple, just include the missing variable or factor in the model fitting function (`lm` or `aov`). However, be sure to refer to the literature for approved practices on when this is permissible and how to go about ensuring that proper experiment design protocol is followed. Recall that indicators of missing variables include unwanted behavior or trends in residual plots and suggestions of mixed distributions in QQ normal probability plots.

15.4   Issues Specific to Covariates

Two possible issues might arise when covariates are included in a model: In the case where two or more covariates are included, there is the possibility that multicollinearity might be present among the covariates; it may also be the case that the response variable does not exhibit an approximate linear relationship with the covariate.

15.4.1   Multicollinearity

Built in the diagnostic procedures described in Sections 9.3 and 9.6 are methods by which the source of multicollinearity can be identified. While a simple solution to eliminating the presence of multicollinearity might be to remove offending covariates of least importance, there are occasions when this may not be advisable. Section 10.7 outlines some suggested alternatives to removing variables that exhibit multicollinearity.

15.4.2   Transformations of covariates

If a nonlinear relationship between the covariate and the response is apparent, the issue can be thought of as a matter of needing to improve fit. As demonstrated in Section 10.2, issues of fit can sometimes be addressed through a transformation of the response variable or through a transformation of the covariate. Methods that might be used here are as for regression models.

15.4.3 Blocking as an alternative to covariates

Consider an observational study involving a one-factor fixed-effects model with a covariate for which it is known that the response is not linearly related to the covariate. During the design phase, the range of the covariate can be partitioned into blocks and the resulting levels can be used in place of the covariate. The rationale for using a block design might be that the requirement of linearity, applicable to covariates, is not applicable to block designs.

The data used in the following illustration are stored in the data frame CovToBlock and saved in the file Data15x01.R (see Table 15.1).

TABLE 15.1: Dataset to Illustrate Changing a Covariate into a Blocking Variable

\multicolumn{2}{c}{Tr1}		Tr2		Tr3	
Y	X	Y	X	Y	X
17.17	1.19	16.28	0.63	11.63	0.18
17.09	1.66	20.28	1.62	21.23	0.96
17.20	1.96	20.20	1.98	17.46	0.54
17.20	1.95	20.30	1.76	10.37	0.13
17.60	1.83	15.93	0.58	22.20	1.89
8.44	0.27	18.05	0.85	22.15	1.23
14.67	0.67	19.18	1.08	11.07	0.19
9.54	0.31	20.01	1.94	19.18	0.65
17.70	1.95	20.07	1.59	22.08	1.15
10.01	0.34	15.20	0.52	22.31	1.45

A command of the form

```
source("z:/Docs/RCompanion/Chapter15/Data/Data15x01.R")
```

loads the data in the workspace, and

```
win.graph(width=3,height=3.5);par(ps=10,cex=.75)
plot(y~x,CovToBlock,pch=unclass(Tr),main=NULL,las=1)
```

produces Figure 15.1, which indicates the response variable is not linearly related to the covariate.

Creating groups for very large datasets is a simple matter with R if the desired partition is known.[1] In Section 3.7, the cut function was used to create

---

[1] The plan here was to simulate a complete block design. Therefore, a little cheating took place; the scatter plot was used to determine an appropriate partition. In practice, this might be frowned upon.

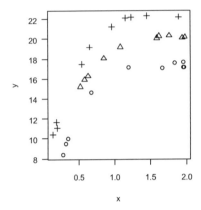

FIGURE 15.1: Scatterplot of dataset `CovToBlock` by factor levels showing a clear nonlinear relationship between y and x.

a grouped frequency distribution. Here, the function `cut` is used to partition the covariate into groups that will serve as blocks.

TABLE 15.2:   Table 15.1 Data Reformatted with a Blocking Variable

X	Tr1	Tr2	Tr3
	8.44	15.93	11.63
[0.0, 0.6)	9.54	15.20	17.46
	10.01		10.37
			11.07
	17.17	16.28	21.23
[0.6, 1.4)	14.67	18.05	22.15
		19.18	19.18
			22.08
	17.09	20.28	22.20
[1.4, 2.0]	17.20	20.20	22.31
	17.20	20.30	
	17.60	20.01	
	17.70	20.07	

Table 15.2 contains the results obtained using the following code:

```
#Determine the lower and upperbounds for the covariate
with(CovToBlock,range(x))
```

```
#Define the partition
bpts<-c(0,0.6,1.4,2)
#Create the classes
blocks<-with(CovToBlock,cut(x,breaks=bpts,
                       include.lowest=T,right=F))
#Store the results
BlockData<-with(CovToBlock,data.frame(y,Tr,blocks))
```

The subsequent analyses are then as in Chapter 13.

15.5   For the Curious

The "observed" data in Table 15.1 were generated using the following code:

```
#Define the mean response function
f<-function(z){15-(z-2)^4}
#Set cell size, initialize variables and "observation" counter
n<-10; Tr<-integer(3*n); y<-numeric(3*n)
x<-numeric(3*n); k<-1
#Start outer for-loop to generate data
for (j in 1:3){
    #Get some x-values and set a treatment mean
    xx<-round(runif(n,0,2),2); mu<-2.5*j
    #Get some random error terms
    e<-round(rnorm(n,mean=mu,sd=.25),1)
    #Start inner for-loop
    for (i in 1:n){
        #Assign factor levels and obtain "observed" responses
        Tr[k]<-j; y[k]<-round(f(xx[[i]])+e[i],2)
        #Save x-values and update "observation" counter
        x[k]<-xx[i]; k<-k+1}}
#Store the data
CovToBlock<-data.frame(y,x,Tr)
#Define Tr as a factor
CovToBlock$Tr<-factor(CovToBlock$Tr)
```

Bibliography

[1] Anscombe, F. J., Rejection of outliers, *Technometrics*, 2(2), 123–147, 1960.

[2] Atkinson, A. C., *Plots, Transformations, and Regression*, Clarendon Press, Oxford, 1985.

[3] Bain, L. J. and M. Engelhardt, *Introduction to Probability and Mathematical Statistics*, 2nd ed., Duxbury Press, Belmont, CA, 1992.

[4] Barnett, V. and T. Lewis, *Outliers in Statistical Data*, 3rd ed., John Wiley & Sons, New York, 1994.

[5] Belsley, D. A., E. Kuh, and R. E. Welsch, *Regression Diagnostics: Identifying Influential Data and Sources of Collinearity*, John Wiley & Sons, New York, 1980.

[6] Bluman, A. G., *Elementary Statistics: A Step by Step Approach, A Brief Version*, 5th ed., McGraw Hill, New York, 2004.

[7] Box, G. E. P., W. G. Hunter, and J. S. Hunter, *Statistics for Experimenters: An Introduction to Design, Data Analysis and Model Building*, John Wiley & Sons, New York, 1978.

[8] Braun, W. J., *MPV: Data Sets from Montgomery, Peck and Vining's Book*. R package version 1.25, 2008. http://cran.r-project.org/web/packages/MPV/.

[9] Braun, W. J. and D. J. Murdoch, *A First Course in Statistical Programming with R*, Cambridge University Press, Cambridge, 2007.

[10] Brillinger, D. R., *Time series: Data analysis and theory*, SIAM, Philadelphia, PA, 1981.

[11] Carroll, R. J. and D. Ruppert, Transformations in regression: A robust analysis, *Technometrics*, 27(1), 1–12, 1985.

[12] Casella, G. and R. L. Berger, *Statistical Inference*, Wadsworth & Brooks/Cole, Pacific Grove, CA, 1990.

[13] Chambers, J. M., W. S. Cleveland, B. Kleiner, and P. A. Tukey, *Graphical Methods for Data Analysis*, Duxbury Press, Belmont, CA, 1983.

[14] Cowpertwait, P. S. P. and A. V. Metcalfe, *Introductory Time Series with R*, Springer, New York, 2009.

[15] Crawley, M. J., *Statistics: An Introduction Using R*, John Wiley & Sons, New York, 2005.

[16] Cryer, J. D. and K. Chan, *Time Series Analysis: With Applications in R*, 2nd ed., Springer, New York, 2008.

[17] Dalgaard, P., *Introductory Statistics with R*, Springer, New York, 2002.

[18] Dixon, W. J., Analysis of extreme values, *Annals of Mathematical Statistics*, 21(4), 488–507, 1950.

[19] Dolby, J. R., A quick method for choosing transformations, *Technometrics*, 5(3), 317–325, 1963.

[20] Draper, N. R. and H. Smith, *Applied Regression Analysis*, John Wiley & Sons, New York, 1998.

[21] Faraway, J. J., *Linear Models with R*, Chapman & Hall/CRC, Boca Raton, FL, 2005.

[22] Faraway, J. J., *Extending the Linear Model with R: Generalized Linear, Mixed Effects and Nonparametric Regression Models*, Chapman & Hall/CRC, Boca Raton, FL, 2006.

[23] Faraway, J., *faraway: Functions and datasets for books by Julian Faraway.* R package version 1.0.4, 2009. http://www.maths.bath.ac.uk/~jjf23/.

[24] Feder, P. I., Graphical techniques in statistical analysis — tools for extracting information from data, *Technometrics*, 16(2), 287–299, 1974.

[25] Filliben, J. J., The probability plot correlation coefficient test for normality, *Technometrics*, 17(1), 111–117, 1975.

[26] Fox, J., *car: Companion to Applied Regression.* R package version 1.2-16, 2009. http://cran.r-project.org/web/packages/car/.

[27] Genz, A., F. Bretz, T. Miwa, X. Mi, F. Leisch, F. Scheipl, and T. Hothorn, *mvtnorm: Multivariate Normal and t Distributions.* R package version 0.9-9, 2010. http://cran.r-project.org/web/packages/mvtnorm/.

[28] Genz, A. and F. Bretz, *Computation of Multivariate Normal and t Probabilities*, Springer-Verlag, Heidelberg, 2009.

[29] George, E. I., The variable selection problem, *Journal of the American Statistical Association*, 95(452), 1304–1308, 2000.

[30] Gilmour, S. G., The interpretation of Mallows' C_p-statistic, *Journal of the Royal Statistical Society*, Series D (The Statistician), 45(1), 49–56, 1996.

[31] Graybill, F. A., *An Introduction to Linear Statistical Models*, McGraw-Hill, New York, 1961.

[32] Graybill, F. A., *Matrices with Applications in Statistics*, 2nd ed., Wadsworth Publishing, Belmont, CA, 1983.

[33] Graybill, F. A., *Theory and Application of the Linear Model*, Duxbury Press, Pacific Grove, CA, 2000.

[34] Green, R. F., Outlier-prone and outlier-resistant distributions, *Journal of the American Statistical Association*, 71(354), 502–505, 1976.

[35] Grubbs, F., Sample criteria for testing outlying observations, *Annals of Mathematical Statistics*, 21(1), 27–58, 1950.

[36] Grubbs, F., Procedures for detecting outlying observations in samples, *Technometrics*, 11(1), 1–19, 1969.

[37] Harrell, F. E. Jr., *Design: Design Package*. R package version 2.3-0, 2009. http://cran.r-project.org/web/packages/Design/.

[38] Hawkins, D. M., *Identification of Outliers*. Chapman & Hall, New York, 1980.

[39] Heiberger, R. M., *HH: Statistical Analysis and Data Display: Heiberger and Holland*. R package version 2.1-32, 2009. http://cran.r-project.org/web/packages/HH/.

[40] Hogg, R. V. and A. T. Craig, *Introduction to Mathematical Statistics*, 4th ed., Macmillan, New York, 1978.

[41] Hothorn, T., F. Bretz, and P. Westfall, Simultaneous inference in general parametric models, *Biometrical Journal*, 50(3), 346–363, 2008.

[42] Johnson, R. A. and D. W. Wichern, *Applied Multivariate Statistical Analysis*, 4th ed., Prentice Hall, Upper Saddle River, NJ, 1998.

[43] Keener, J. P., *Principals of Applied Mathematics: Transformations and Approximation*, Addison-Wesley, New York, 1988.

[44] Koenker, R., *quantreg: Quantile Regression*. R package version 4.53, 2010. http://cran.r-project.org/web/packages/quantreg/.

[45] Kutner, M. H., C. J. Nachtsheim, J. Neter, and W. Li, *Applied Linear Statistical Models*, 5th ed., McGraw-Hill, New York, 2005.

[46] Looney, S. W. and T. R. Gulledge Jr., Use of the correlation coefficient with normal probability plots, *The American Statistician*, 39(1), 75–79, 1985.

[47] Lumley, T., Using Fortran code by Alan Miller, *leaps: Regression Subset Selection.* R package version 2.9, 2009. http://cran.r-project.org/web/packages/leaps/.

[48] Marsden, J. E. and A. J. Tromba, *Vector Calculus,* 3rd ed., W. H. Freeman and Company, New York, 1988.

[49] Mendenhall, W., *A Second Course in Statistics: Regression Analysis,* 6th ed., Pearson Education, Upper Saddle River, NJ, 2003.

[50] Montgomery, D. C., E. A. Peck, and G. Vining, *Introduction to Linear Regression Analysis,* 3rd ed., John Wiley & Sons, New York, 2001.

[51] Montgomery, D. C., *Design and Analysis of Experiments,* 4th ed., John Wiley & Sons, New York, 1997.

[52] Murdoch, D. and E. D. Chow (porting to R by Jesus M. Frias Celayeta), *ellipse: Functions for Drawing Ellipses and Ellipse-like Confidence Regions.* R package version 0.3-5, 2007. http://cran.r-project.org/web/packages/ellipse/.

[53] Myers, R. H., *Classical and Modern Regression with Applications,* Duxbury Press, Boston, MA, 1986.

[54] Oehlert, G. W., *A First Course in Design and Analysis of Experiments,* W. H. Freeman and Company, New York, 2000.

[55] Pinheiro, J., D. Bates, S. DebRoy, D. Sarkar, and the R Core team, *nlme: Linear and Nonlinear Mixed Effects Models,* 2009. http://cran.r-project.org/web/packages/nlme/.

[56] R Development Core Team, *R: A Language and Environment for Statistical Computing.* R Foundation for Statistical Computing, Vienna, Austria, 2010. http://www.R-project.org.

[57] Sarkar, D., *lattice: Lattice Graphics.* R package version 0.18-8, 2010. http://cran.r-project.org/web/packages/lattice/.

[58] Scheffé, H., *The Analysis of Variance,* John Wiley & Sons, New York, 1959.

[59] Searle, S. R., *Matrix Algebra Useful For Statistics,* John Wiley & Sons, New York, 1982.

[60] Searle, S. R., *Linear Models,* John Wiley & Sons, New York, 1971.

[61] Shapiro, S. S. and M. B. Wilk, An analysis of variance test for normality (complete samples), *Biometrika,* 52(4), 591–611, 1965.

[62] Sheskin, D. J., *Handbook of Parametric and Nonparametric Statistical Procedures,* 2nd ed., Chapman & Hall/CRC, Boca Raton, FL, 2000.

[63] Siniksaran, E., A geometric interpretation of Mallows' C_p statistic and an alternative plot in variable selection. *Computational Statistics & Data Analysis*, 52(7), 3459–3467, 2008.

[64] Smith, J. H., Families of transformations for use in regression analysis, *The American Statistician*, 26(3), 59–61, 1972.

[65] Therneau, T. and original R port by Lumley, T., *survival: Survival Analysis, Including Penalised Likelihood.*. R package version 2.35-8, 2009. http://cran.r-project.org/web/packages/survival/.

[66] Thode, H. C. Jr., *Testing for Normality*, Marcel Dekker, New York, 2002.

[67] Timm, N. H., *Applied Multivariate Analysis*, Springer, New York, 2002.

[68] Tukey, J. W., On the comparitive anatomy of transformations, *Annals of Mathematical Statistics*, 28(3), 602–632, 1957.

[69] Tukey, J. W., *Exploratory Data Analysis*. Addison-Wesley, Reading, MA, 1977.

[70] Wehrens, R. and B. Mevik, *pls: Partial Least Squares Regression (PLSR) and Principal Component Regression (PCR)*. R package version 2.1-0, 2007. http://mevik.net/work/software/pls.html.

[71] Zeileis, A. and T. Hothorn, Diagnostic checking in regression relationships. *R News* 2(3), 7–10, 2002. http://cran.r-project.org/doc/Rnews/.

Index

Adjusted coefficient of multiple determination, 173, 219
Arithmetic operations, 24
Assignment operator, 36
Autoregressive methods, 243

Balanced design, 252
Baseline treatment, 255
Binary response data, 243
Binding two or more vectors, 37
Block designs, 301
Block model
    Additive effects, 302
    Blocking effects, testing, 306
    Cell means, 301
    Interaction effects, testing, 305
    Main effects, testing, 306
    Nonadditive effects, 302
    Treatment coded, additive, 303
    Treatment coded, nonadditive, 302
Bonferroni's outlier test, 113, 115
Boolean, 35
Boolean algebra, 85
Box–Cox procedure, 141, 163
Box–Tidwell procedure, 233, 243
Boxplots, 61

Categorical data, 35
Cautions
    anova function and regression models, 105, 171
    Attaching an object more than once, 17
    Attaching more than one object at a time, 17
    Case sensitivity in R, 5

Data type classification of list contents, 39
Graphics add-ons and what to watch for, 77
Heirarchy for list contents, 39
Incorrect computations, 26
Loading more than one package at a time, 14
Numeric variable names, 38
Outer product and vector multiplication, 47
Reserved letters T and F, 84
Roots of negative numbers, 26
Spaces in variable names, 38
User-defined functions and the library function, 34
Cell means, 301
Coding scheme
    Contrast for, 255
    Resetting, 255
    Sum contrast, 270
    Treatment coding, 255
    Weighted sum contrast, 268
Coefficient of determination, 105
Coefficient of multiple determination, 173, 218
Combinations, 25, 262
Condition indices, 225
Condition number, 224
Conditional statements
    If-stop, 87
    If-then, 88
    If-then-else, 88
    ifelse function, 89, 92
Constant variance test
    Brown–Forsyth test, 110

F-test to compare two variances, 108
Constants, 25
Continuous piecewise regression, 151
Contrast for general hypothesis tests, 265
Correlation coefficient matrix, 55
Correlation transformation, 222
Covariance model
    Centered, 284, 299
    Constant slopes assumption, 287
    Treatment effects, 288
    Uncentered, 283
Covariates, 283

Data formats in R
    Lists vs. data frames, 176
    Multivariate numeric, 39
    Multivariate with factors, 39
    Univariate, 39
Data structure, 35
Datasets used
    AncovaData, 284
    Block2Data, 312
    BlockData, 303
    BoxTidData, 233
    CatReg1, 156
    CatReg2, 160
    CatSet, 49
    lake.data, 37
    MultiNum, 54
    MultipleReg, 168
    OneWayData, 252
    PolyReg, 147
    Retreat, 74
    RightSkewed, 139
    Salmon, 60
    SimpleRegData, 102
    SmallSet, 47
    ThreePiece, 154
    TwoPiece, 150
    TwoWayAdd, 321
    UniVarSet, 53
    XTransReg, 134
    XYTransData, 235

XYTransReg, 137
Decimal fractions, 25
Deleted model, 182
Discrete response data, 243
Dot product, 45
Durbin–Watson test, 113

Effects parameters, 302, 319
Eigenpair, 223
Eigenvalue, 223
Eigenvector, 223
Exporting output to text file, 57
Extra sum of squares principle, 211

Factor, 251
Factor levels, 251
Factorials, 25
File management
    Directory, 3
    Drive on a computer, 3
    Folder, 3
    Path, 4
    Start-in directory, 6
    Sub-directory, 3
    Working directory, 4

Generalized least squares regression, 242
Graphics
    Copying a graphics object, 17
    Customizing graphics parameters, 61
    Customizing window and point size, 61
    Enhancing plots, 77
    Identify points on a scatter plot, 114
    Line segments, 151
    Line widths, 128
    Line/curve colors, 125
    Mathematical annotation, 165
    Multiple plots in one window, 65
    Plot symbols, 128
    Saving a graphics object, 17

Half-normal plot, 73, 79

Hat-matrix, 173
Histogram
    Density, 63
    Frequency, 63
    With normal curve, 67

Importing datasets, 47
Influence measures
    Cook's distance, 182, 184
    COVRATIO, 184, 185
    DFBETAS, 182, 184
    DFFITS, 181, 184
Inhibit function, 136
Inner product, 45
Interaction plot, 303

Knot points, 153

Lack of fit test, 205
Least significant difference test, 262
Leverages, 173
Loading source code and other objects, 12
Logical
    Data, 35
    Initializing variables, 84
    Operations, 83
    Operator, 84, 85
    Statements, 84, 85
    Values, 83
    Variables, 83
Logistic regression, 243
Loops
    Entry condition, 91
    Exit condition, 90

Mac alert
    Differences from Windows, 6
    Graphics window, 61
Matrix multiplication, 46
Matrix scatterplots, 76
Mean square error, MSE, 105
Mean treatment effect, 252
Mean treatment response, 252
Messages from R
    Computational warnings, 26

Detaching a package, 16
Detaching multiple packages, 15
help command, 8
help.search command, 9
Masked objects, 17
NA, 28
NaN, 26
Object not found, 36
Outdated packages, 16
Model selection criteria
    Adjusted coefficient of multiple determination, 219
    Akaike information criterion, 213, 220
    Bayesian information criterion, 220
    Biased prediction, 219
    Biased regression coefficient estimates, 219
    Coefficient of multiple determination, 218
    Mallow's statistic, 219
    Mean square error, 218
    Overspecified models, 219
    PRESS statistic, 221
    Schwarz–Baysian criterion, 220
    Underspecified models, 219
Modular arithmetic, 25
Multicollinearity diagnostics
    Correlation coefficient, 209
    Correlation coefficient matrix, 209
    Variance inflation factor, 209
    Variance inflation factors, 210
Multiple regression model, 168
Multivariate data
    Assigning or reassigning levels to categorical variables, 50
    Data frames, 49
    Defining a variable as a factor, 50
    Descriptive statistics, 54
    Extracting rows by properties, 50
    Extracting specific columns, 51
    Importing from external files, 49
    Matrix operations, 51

Saving a data frame, 51

Naming data structures, 36
Nonparametric methods, 243
Normality test
    QQ normal probability correlation coefficient test, 111
    Shapiro–Wilk test, 112
Numbers in R
    Decimal form, 24
    Integers, 23
    Irrational numbers, 23
    Rational approximations, 24
    Rational numbers, 23
    Real numbers, 23
    Scientific notation, 24
    Testing functions, 25
Numeric data, 35

Objects in R
    Activation of, 36
    Assignment of values, 36
    Attaching an object, 16
    Combining objects, 37
    Data frame, 37
    Definition of, 6
    Initializing, 38
    Listing attached objects, 17
    Listing objects, 11
    Listing sub-objects, 16
    Mode type, 38
    Reassigning of names, 37
    Renaming levels in a factor, 60
    Renaming variables in a data frame, 60
    Saving a data frame, 48
One-factor model, 252
Outer product, 46
Overall mean response, 252

Packages
    Citing R packages, 19
    Default loaded, 12
    Definition of, 12
    Detaching a loaded package, 15
    Detaching multiple packages, 15

Installed, 13
Installing packages from CRAN, 13
Listing installed packages, 13
Listing loaded packages, 12
Loading an installed package, 14
Masked objects, 14
Partial F-test, 211
Partial least squares regression, 242
Piecewise regression, 149
Plotting position
    Half-normal plots, 79
    QQ normal probability plots, 78
Poisson regression, 243
Polynomial regression, 147
PRESS residuals, 221
Principal components, 225
Principal components analysis, 243
Principal components regression, 242
Probability functions
    F-, 31
    Normal distribution, 29
    Studentized range distribution, 275
    t-, 30
Programming in R
    Case sensitivity, 5
    Comments, 93
    Executing commands in command line environment, 5
    Syntax for R commands, 5
    Use of semicolons in command lines, 11

QQ lines, 71
QQ normal probability plot, 70, 78
    Interpreting, 81
Quantiles
    F-, 32
    Normal, 30
    Studentized range distribution, 275
    t-, 31
    Uniform, 134

R console

Clearing the console, 7
Command line, 5
Editing command lines, 7
Exiting R, 6
R script editor, 20
R syntax, 5
!, 84
!=, 85
+, -, *, /, ^, 24
.( ), 96
:, 27
==, 85
[...], 41
[[...]], 41
#, 93
$, 41, 48
&, 84
<, 85
<-, 36
<=, 85
>, 85
>=, 85
|, 84
abline, 71, 78
abs, 24
add1, 213
ancova, 288
anova, 157
aov, 256
ar, 243
as.matrix, 45, 46
as.vector, 44
asin, 24
attach, 16, 48
axis, 78
box.tidwell, 232
boxcox, 142
bquote, 96
by, 253
C, 269
c, 40
cbind, 37
ceiling, 25
cex, 69
character, 38

choose, 25
ci.int, 130
citation, 19
class, 110
confint, 117, 186, 272
contr.sum, 270
contr.treatment, 255
cooks.distance, 182
cor, 55
cov, 55
crossprod, 46
data.frame, 37
deparse, 96
detach, 15, 48
df, 31
dfbetas, 182
dffits, 182
dim, 53
dnorm, 29
drop1, 213
dt, 30
dump, 11
dwtest, 113
e, 25
eigen, 209
ellipse, 197
else, 88
exp, 24
expression, 141
factor, 50, 59, 253
factorial operator, 25
fitted.values, 106
fivenum, 53
floor, 25
for, 90
function, 32
glht, 265
gls, 242
halfnorm, 73
hatvalues, 174
help, 8, 9
help.search, 7
hist, 63
history, 18
hov, 257

I, 136
identify, 114
if, 88
ifelse, 89, 92
influence.measures, 184
installed.packages, 13
integer, 38
interaction.plot, 303
is.integer, 25
is.real, 25
isTRUE, 84
jitter, 290
legend, 78
length, 41, 53
levels, 50, 59
library, 9, 13, 14, 73
lines, 68, 78
list, 176
lm, 104, 171, 209
lm.ridge, 243
load, 6
log, 24
logical, 38, 84
lrm, 243
ls, 11
ltsreq, 242
matplot, 124, 128
max, 53
mean, 53
median, 53
mfrow, 65
min, 53
mode, 25
modulus operator, 25
mtext, 78, 96
names, 16, 37, 59
numeric, 38
options, 255
order, 42
outlier.test, 115
pairs, 76
pairwise.t.test, 262
par, 61, 69, 78
paste, 93, 96, 141
pcr, 242

pf, 31
pi, 25
plot, 75
plot.ts, 74
plotmath, 77
plsr, 242
pnorm, 29
points, 124, 128
predict, 120, 189
PRESS, 222
prod, 43
pt, 30
ptukey, 275
q, 6
qf, 31
qnorm, 29
qqline, 110
qqnorm, 70, 71, 110
qt, 30
qtukey, 275
quantile, 53
quotient operator, 25
range, 53
rank, 42
rbind, 37, 266
read.csv, 47
read.table, 47
rep, 59
require, 14
residuals, 106
rf, 31
rlm, 242
rm, 11, 12
rnorm, 29
round, 25
rq, 242
rstudent, 106, 174
rt, 30
runif, 134
sample, 28
save.image, 6
savehistory, 19
sd, 53
search, 12
segments, 151

seq, 27
setwd, 5
shapiro.test, 112
sign, 25
signif, 25
sink, 57
solve, 52, 209
sort, 41
source, 12
split, 166
sqrt, 24
step, 214
stop, 87
stripchart, 69, 253
substitute, 96
sum, 28, 43
summary, 53, 60, 105
summary.lm, 256
t, 45
tapply, 60, 253
text, 78
title, 78
trunc, 25
ts, 74
TukeyHSD, 263
unclass, 156
var, 53, 55
var.test, 108
vector, 38
vif, 209
while, 91
win.graph, 61
with, 48, 60, 253
xor, 84
Random sample
    Binomial, 91, 207, 316
    F-distributed, 32
    Normally distributed, 30
    t-distributed, 31
    Uniform, 134, 207, 316
Regression sum of squares, 105
Regression with categorical variables
    Non-parallel straight line regression, 159

Parallel straight line regression, 156
Relational symbols, 85
Residual
    Deleted , 221
    Standard error, 105
    Standardized, 106, 173
    Studentized, 173
    Studentized deleted, 106, 173
    Sum of squares, SSE, 105
    Unstandardized, 106, 173
Ridge regression, 243
Robust regression methods, 242

Saving objects, 11
Scalar, 35
Scatterplots, 75
Scientific notation, 25
Searching for help
    Citing R, 19
    Citing R packages, 19
    Function or operators, 8
    Help menu, 7
    Keyword searches, 7
    Package, brief information, 9
    Packages, detailed information, 9
Sequences
    Arithmetic, 27
    Colon operator, 27
    Consecutive integers, 27
    Geometric, 28
    Random sequences/samples, 28
    Regular sequences with step-size one, 27
    Sum of, 28
    Using mathematical functions, 28
Session history
    Saving a full R session, 19
    Saving to the default file, 19
    Viewing recent code, 18
Simple regression model, 102
String, 35
Stripchart, 69
Subset regression, 150

Sum of squares of total variation, SSTo, 105
Summaries by factors, 60

Time-series, 73
Tinn-R script editor, 97
Treatment cell, 252
Treatments, 251
Tukey's nonadditivity test, 312
Tukey–Kramer procedure, 275
Two-factor model
    Additive effects, 320
    Cell means, 319
    Interaction effects, testing, 323
    Main effects, testing, 324
    Nonadditive effects, 320
    Scheffes's procedure, 331

Unbalanced design, 252
Univariate data
    Accessing entries/contents, 41
    Coercing a list into matrix format, 45
    Coercing a list into vector format, 44
    Computations with lists of equal length, 42
    Computations with lists of unequal length, 43
    Create with the combine function, 40
    Descriptive statistics, 53
    Ordered index list, 42
    Ranks of entries, 42
    Sorting entries, 41
    Special vector operations, 46
    Transposing a vector, 45
Unweighted mean, 252

Variance proportions, 226
Variance-covariance matrix, 55
Vector, 35

Weighted least squares regression, 242
Weighted mean, 252, 268
Windows in R

Graphics device, 6
R Console, 5
Script editor, 5
Switching R windows, 7
Working directory in R
    Current working directory, 5
    Setting the working directory, 4
    Start-in directory, 4
Workspace
    Book-keeping, 10
    Default workspace file, 6
    Definition of, 6
    Listing objects, 11
    Loading a workspace, 6
    Loading source code/objects, 12
    Removing all objects, 12
    Removing objects, 11
    Saving a workspace, 6

Printed and bound by CPI Group (UK) Ltd, Croydon, CR0 4YY

24/10/2024

01778595-0001